DON'T LOOK, DON'T TOUCH, DON'T EAT

Valerie Curtis

DON'T LOOK,
DON'T TOUCH,
DON'T EAT

The Science
Behind Revulsion

The University of Chicago Press
Chicago and London

VALERIE CURTIS is director of the Hygiene Centre
at the London School of Hygiene and Tropical
Medicine.

The University of Chicago Press, Chicago 60637
The University of Chicago Press, Ltd., London
© 2013 by The University of Chicago
All rights reserved. Published 2013.
Printed in the United States of America

22 21 20 19 18 17 16 15 14 13   1 2 3 4 5

ISBN-13: 978-0-226-13133-7 (cloth)
ISBN-13: 978-0-226-08910-2 (e-book)

DOI: 10.7208/chicago/9780226089102.001.0001

Library of Congress Cataloging-in-Publication Data

Curtis, Valerie, author.
    Don't look, don't touch, don't eat: the science
behind revulsion / Valerie Curtis.
        pages; cm
    ISBN 978-0-226-13133-7 (cloth: alkaline paper) —
ISBN 978-0-226-08910-2 (e-book)   1. Aversion.
2. Aversion—Social aspects.   3. Hygiene—
Psychological aspects. I. Title.
    BF575.A886C87 2013
    152.4—dc23   2013017713

CONTENTS

PREFACE   UNWEAVING THE RAINBOW

Feces, urine, toilets, sweat, menstrual blood, spilt blood, cut hair, impurities of childbirth, vomit, smell of urine, open wound, saliva, dirty feet, eating with dirty hands, food cooked while menstruating, bad breath, smelly person, yellow teeth, nose picking, dirty nails, clothes that have been worn, flies, maggots, lice, mice, mouse in a curry, rats, stray dog, meat, fish, pigs, fish smell, dog or cat saliva, flies on feces, liquid animal dung, soap that has been used in the latrine, dead rat, rotting flesh, parasitized meat, wet cloths, stickiness, decaying waste, garbage dump, sick person, hospital waiting rooms, beggars, touching someone of lower caste, crowded trains, alcohol, nudity, kissing in public, betrayal.

From fifty essays on disgust by teenage girls in Lucknow, India

Read this collection of the foul, the fetid, and the revolting carefully. Savor each word; imagine each object. How does this make you feel? The list comes from a set of essays on the subject of disgust, written for me by teenage girls in the city of Lucknow. I happened to be having breakfast while I was working on it. Partway through I had to put my spoon down; by the last essay my stomach was churning. Perhaps you feel the same sensations that I did: a curdling of the tongue, a clamminess of the hands, a slowing of the heartbeat? Maybe you also pulled down the sides of your mouth and wrinkled your nose in the classic facial expression of disgust. If you were dining while reading this list, as I was, you will probably also have set aside your plate.

Disgust's Puzzle

Why should this simple collection of words on a page produce so visceral a reaction? What is there in this list of the odorous, the grubby, the leftover, the discarded, and the immoral that is so powerful it can leap from a page to contaminate a bowl of cereal? What links all of these disparate phenomena—from sweat, to lice, to sick people, to betrayal—apart from the fact that they disgust us? And what is the purpose of this unpleasant feeling—what good does it do us?

All humans, from India to Africa to the USA, feel disgust, and, with a few exceptions, all humans feel disgusted about more or less the same things. We are disgusted not just by particular nasty organic objects and substances, but also by hygienic lapses, and by indecency, wickedness, and hypocrisy. We also seem to suffer from what might be called a "metaphysical disgust," an aversion to violations of the things that we hold sacred, our social rules, and our laws.

Disgust pushes us away; it is a voice in the head telling us to avoid the nasty, the stinky, the putrid, and the offensive. Yet at the same time it fascinates. Freak shows at fairs have always been a draw, as have been body parts in jars. The news industry thrives on our desire for gory and ghastly details, and bloodsucking vampires, slimy aliens, and deformed monsters fuel the horror fiction and film industries. Some contemporary artists like to add dung to their works, and the more disgusting the tasks that are given to celebrities in TV game shows, the higher the ratings climb. It seems that we like to take our emotions out for a spin; we can't help taking a second look at something that makes us squirm.

Disgust has always been fascinating, but it has only recently become an object of serious study. There is now a miniboom in books, papers, and conferences on pollution and purity, disgust and dirt. Classicists, historians, humanities students, anthropologists, cultural and legal scholars, bioethicists, philosophers, marketers, and psychologists are all wrestling with this previously taboo topic. The total number of books on disgust has doubled since 2011, with four new volumes appearing with titles such as *Yuck!* and *That's Disgusting!*[1] Even the *New York Times* noticed that disgust is a hot topic.[2]

From this burgeoning effort to peer into the nether regions of our lives, a new body of knowledge is beginning to emerge. We now know, for example, that disgust activates particular parts of our brains; that disgust informs our attitudes to others, particularly the sick and the different; that it affects our attitudes toward cloning and to GM foods, our judicial systems, our politics; and that it even affects our product-buying behavior. The previously outré topic of toilets is a new subject of academic study for archaeologists,[3] and the problem of how to get sanitation to the 40 percent of people on the planet who still lack a latrine is, at last, getting serious attention from international development agencies such as the World Bank and UNICEF. Psychotherapists have learned that disgust can trump love; once disgust of the nasty habits of one's partner takes over, a relationship is likely to be doomed.[4] Social psychologists have reported the results of experiments that show disgust in operation when humans make moral choices. Disgust, it turns out, is a thread that is woven right through our individual and social lives; indeed, without this thread societies would fall apart.

Sorting Out the Disgust Story

However, while there is now a marvelous flowering of work on disgust, the topic does not yet have solid roots. Disgust studies grow in a motley compost of observation, out-of-date theory, and speculation. Some students think disgust emerges out of social norms and culture, others find magical thinking about contamination at its roots, others claim that it grows out of an innate distaste for bad food, and many subscribe to the view that it serves to defend our psyches—to keep us from too much awareness of the unpleasant facts of our animal nature and of our ultimate destiny, which is to become organic waste material ourselves.[5] One of the most influential theories puts disgust of dirt at the heart of the way societies are ordered. The argument goes that if we did not reject and revile certain objects and actions, they would threaten the whole fabric of society.[6] Some theorists have even thrown their hands up and suggested that the manifestations of disgust are so multiform and so various that they defy coherent explanation.[7]

Disgust certainly is puzzling; it is so powerful and sometimes so apparently irrational that it can seem magical. To find it playing so many roles in so many spheres of life seems to defeat scientific explanation. Yet science and the scientific approach *can* help us make sense of this messy and multifarious topic. We can systematically trace the ancient origins of disgust back to our earliest animal ancestors. We can hypothesize about disgust's functions and its component mental structures and test these hypotheses in rigorous experiments. We can unpick the story of how and why disgust plays a role in our moral behavior. And we can use it to unweave the rainbow:[8] to find the strands that can make sense of age-old questions such as how our ancient biological natures have influenced the content of our cultures, and how our cultures have, in turn, influenced our natures.

The story of disgust needs to be retold, this time from the start, and in the right order. Therefore this book starts with disgust's ancient function as a system that bestows on animals the ability to avoid parasites—those ubiquitous body snatchers that hitch lifts and sneak free lunches from their hosts. As the story unfolds, we see how invertebrates, vertebrates, mammals, and primates evolved the behavioral capacities needed to deal with threats from parasites and pathogens.[9] Parasite-avoidance behavior is everywhere in the evolutionary tree of life; humans are but one branch, with more in common with our animal cousins than we may like to think.

Humans, however, have also developed some special capacities that are beyond other animals. We are probably the only species that can track parasites by using our imaginations—by conjuring pictures of spreading waves of contagion in our minds. We are probably also the only species to be able to imagine ourselves being disgusted in the future. This has huge evolutionary benefits—we can stay home when there's plague about to avoid encountering sick people. We can choose to put food in the refrigerator today because we can imagine a rotten mess tomorrow. Humans are also the only species with manners—sets of behavioral rules that protect one another, the absence of which occasions disgust and the punishment of the offender. Finally, humans have found a novel use for disgust: employing it to punish social parasites—shunning and excluding the thief, the abuser, and the cheat from society. Recent experiments suggest that disgust may have played a fundamental role in the evolution of morality. Moral disgust has a strong claim to be essential to the human ability to cooperate on a mass scale, which is the main reason that *Homo sapiens* is such an exceptional and successful kind of animal.

It is important that we better understand the biological roots of disgust, not just because disgust underpins morality and altruism, but also because it fuels some of the worst behavior in our societies. Disgust lurks on the dark side of human nature, rearing its ugly head in bullying; cruelty; class hatred; the exclusion of the sick, the aged, and the disabled; homophobia; racism; war; and genocide. Disgust can all too easily be turned on others and used as a weapon. Disgust teaches us wider lessons too; it tells us something about the nature of emotions—where they come from, what they are for, how they work, and how they shape our behavior as individuals and in groups. Disgust has an irresistible story to tell about what it means to be human.

Living with Disgust

I've been on the trail of disgust for many years. I itch when I think about my trip down the Cloaca Maxima—the hot, stinky, and extremely claustrophobic ancient sewer that carries Rome's wastes into the river Tiber. Disgust has invaded every part of my life. Dinner-party conversation somehow inevitably turns to speculation about the revolting, usually as an accompaniment to dessert. "This is Val," my brother says when introducing me, "she's big in poo." And in my turn, I have given friends, colleagues, and students plenty to laugh about with the plastic turd I carry around and the nasty stories I've collected.

Disgust is great fun, but it is also deadly serious. In Africa two-thirds of all deaths are due to infections. In the UK the infection death rate is only one in twenty. My experience of living a large part of my professional life in Africa taught me that this scandalous disparity can be resolved, and that investment in infection prevention is money well spent. Living in Africa also taught me that infections can be so common that the parasites that cause them must have exerted a strong selection pressure on brains and behavior. While carrying out studies of hygiene behavior in Bangladesh, Burkina Faso, China, India, Uganda, Vietnam, Indonesia, and Kyrgyzstan, I saw that different rules of behavior around cleaning and eating seemed to have a common thread—that of courtesy with bodily fluids, of not disgusting others, if at all possible. Far from being a superficial issue, I saw that manners are an important way for people to show that they are conforming to society's rules and thus avoid social exclusion. As I followed the genocide in Rwanda, I saw the vocabulary of disgust exploited with hypocrisy by leaders who knew how to use it to bolster their own power.

In short, I realized that disgust has a powerful hold over us, dictating what we do, not just about microbes but also about manners and morality. So now, instead of turning away in revulsion, dear reader, I invite you to look, touch, and eat disgust (at least in imagination). You are invited on a journey of discovery, into the dark side of human nature.

CHAPTER ONE

EVASION OF THE BODY SNATCHERS

Pus, vomit, urination, menstruation, sexual fluids and so on—all substances and acts that, for some reason, many cultures tend to see as repellent and, despite their constant presence in human life, as abnormal.

A. K. Reinhart, "Impurity/No Danger"

Eureka moments are what every scientist lives for. I'd been mired in sheaves of data, mostly long transcripts of interviews with women in India, Africa, and Europe about their personal hygiene. I'd been asking them what they did, why they cleaned up after their kids, why they used soap and detergents, why they groomed themselves and tidied their houses, and why they washed their hands, or—more often—why they didn't. The women found it surprisingly hard to explain their cleaning and preening behavior. Often as not, they said that they felt that they needed to remove things that were "yuck"—the smelly, the clammy, the ugly, or the nasty. But when I asked them to explain why, we got stuck. "They are just yuck!" they would say.

Here are some of the things that Dutch women found revolting: feces, cats, aphids in lettuce, hairs, dogs, pollution, vermin, dog hairs, drug users, vomit, dog saliva, drunkenness, dust, rotten waste, fat people, sweat, bad-smelling food, insulting, stickiness, food leftovers, politicians, offal, worms, dirty old men, fishmongers' hands, flies.

And here is what women in Burkina Faso told me that they found disgusting: feces, dirty latrine, dirty food, unswept yard, diarrhea, impure substances associated with birth, flies on food, sores, rubbish in the yard, worms, sexual relations before a child is weaned, smelly drains, dirty clothes, rubbish heaps, sick people, pigs, vomit.

Closer to home, I asked a group of women in Cheshire, in the north of England, to write about disgust, and these are what they chose: feces, stained kitchen, dirty hotel, dog shit, flies, dirty cafeteria, cat shit, dead sparrow, dirty play area, dog diarrhea, rotten food, drunks, child shit on the sofa, moldy food, drunken louts, vomit, rank smell of old grease, rude people, wounds, dirty nails, foul language, maggots, eating with mouth open, man beating a woman, sweaty person, eating a burger that a stranger had bitten into, body parts in jars, cruelty to a horse, stained toilet, cleaning another's false teeth, wounding of an old lady.

If you compare these Dutch, African, and English examples with the Indian list in the introduction, you'll see that what, at first sight, looks like a ragbag of unrelated rejecta does, in fact, have themes. All of the lists include bodily wastes and body products. All include bad foods, contaminated objects, and immoral behavior. But, like the women who had participated in these studies, I was also stuck for an explanation. What could possibly account for the fact that such disparate items all occasioned a similar response—a feeling of "eugh!," accompanied by a wish to push them all away, to avoid them, to remove them—and what could explain my desire, even as I was poring and puzzling over them, not to think about them at all?

Then came the eureka moment. I'd been asked for advice about a rare disease caused by a lung fluke in Asia. I wasn't certain how to answer the question, but I knew where to go—to the standard reference work on infectious disease: the *Control of Communicable Diseases Manual*.[1] I flipped through the index, and suddenly a pattern jumped out at me. Interspersed between diseases such as acanthamebiasis and zygomycosis came familiar words: "contaminated food," "dog faeces," "flies," "hairs," "human faeces," "pigs," "rubbish," "sexual fluids," "vomit," "worms," and "wounds." All topics from my disgust lists. There was clearly a pattern here.

I went back to the archive of internationally disgusting items, and, sure enough, it seemed that almost everything there could be found in my infectious-disease compendium.

Some of the most disgusting bodily products turned out to be the most deadly. Feces are not just revolting; they are the source of over twenty gastroenteric infections, including cholera, typhoid, cryptosporidiosis, rotavirus, and the other stomach bugs that are responsible for three-quarters of a million child deaths a year.[2] Nasal mucus is not just nasty: it carries the agents of tuberculosis, influenza, measles, leprosy, and the pneumonias that kill even more children annually. Saliva transmits herpes, syphilis, and mumps. Blood carries AIDS; hepatitis B, C, and D; Lassa fever; syphilis; and trypanosomiasis. Sexual fluids transmit chlamydia, herpes, AIDS, gonorrhea, syphilis, and trichomoniasis. Other bodily fluids can pass on impetigo, chicken pox, smallpox, diphtheria, thrush, ringworm, influenza, leprosy, meningitis, and German measles.

And the index contained not just the disgusting body products of humans—disgusting animals were also there. The rat carries plague and a variety of interesting parasites such as those that cause arenaviral hemorrhagic fever, Lassa fever, and Weil's disease. Snails and slugs carry helminthic parasites. Insects such as flies and cockroaches walk about in wastes, spreading the agents of gastroenteric infection. Other insects, like the louse and the scabies mite, are themselves parasites. Some insects are both parasites and parasite vectors at the same time: fleas carry plague and typhus, lice carry relapsing and trench fever, ticks carry encephalitis and a variety of viral fevers. Earthworms are not dangerous to human health, but they look similar to the parasitic worms that can be found in meat and fish and that infect over a third of humanity.

It seemed that I had stumbled on an explanation for the diversity of the disgusting. All of the things that people find revolting seemed to have some sort of role to play in the transmission of infectious disease.[3] Eureka? Almost—only one problem remained. Why did cruelty, rudeness, drunkenness, and politicians appear on my lists of the disgusting? Moral disgust seemed to need a different explanation. I decided to put this problem aside for the moment and explore the role of disgust in protecting us from infectious disease.

The Birth of PAT

Like any scientist with a hypothesis, I needed to test it with more data. Where better to find data than the international departure lounge of Athens airport? A student of mine asked over 250 people from Europe, the Middle East, the United States, and Africa about what they found disgusting. Here is his list:

Dog feces in the street; feces; dirty baby diapers; animal saliva; sputum; spitting; bogies (nasal mucus); sweat; people who do not brush their teeth; yellow teeth; vomit; seeing someone vomit; smokers; rotten food; rotten things in the refrigerator; rubbish everywhere; dirty hands before a meal and before bed; no oral hygiene; bad smells (body odor, drains, mushrooms cooking); bad body smell; smelly feet; stinky feet; bad breath; passing wind; body odor; picking nose in public; bad breath; if someone doesn't blow his nose but sniffs, especially in public transport; mucus; eaten alive by insects; cockroaches in my bed; millions of bugs clustered together; insects; cockroaches; grasshoppers; snakes; childbirth; all kinds of pigeons and birds; seagulls; dirty people; dying person; eating with mouth open; wet people; wet shoes; dirty dishes; dirty clothes; old cars; seaweed; when someone has a cut finger; London Underground; pollution; rude Europeans; all Americans; Serbian government; politicians; people who eat dogs, snakes, etc.; violent verbal abuse; physical abuse; smoking when children are around; smoking in non-smoking areas; lies; spite; killing animals; animal maltreatment; injustice; [being] ripped off for being a tourist; too much beer; poverty; child abuse; masturbation in public; people who don't want to learn things from others; British society in general; materialism; inequality; teenager getting pregnant to get money from the state; housing allocation for single mums in the inner cities; no tenderness with child; pornography; arrogant people breaking the traffic law; pickpocket; fighting; lack of respect.[4]

Setting aside again the rather interesting moral violations (politicians and Americans got the highest disgust scores in this sample), the pattern held up. Bodily fluids and wastes appeared, as well as foods that were off; the insect, bird, and animal vectors of disease; and poor hygiene, all of which are implicated directly, or indirectly, in infectious-disease transmission. Even seaweed, it turned out when I looked it up, can harbor the vibrion that causes cholera.[5]

So was born the parasite avoidance theory of disgust (or PAT for short). The reason we can't resist wanting to recoil when we meet the nasty, the foul, and the stinky is ancient and instinctual.[6] It's not a reasoned response to "knowing" about germs and disease; rather, it comes from an ancient wellspring of wisdom: the process of evolution. Those of our distant ancestors who tended to avoid feces, nasal mucus, and bad-smelling food did better on average in the reproduction lottery; they were healthier, mated more often, brought up more children to sexual maturity, and hence had more grandchildren. And these grandchildren, the descendants of the disgusted, were more disgustable themselves—and so on, till the present day, and us.

Pathogens and parasites caused the evolution of defense systems.[7] These included not just impermeable skin, toxic secretions, and internal immune systems to kill parasites once they had gotten inside the body, but also behavioral defenses that kept potential hosts away from parasites in the first place. We all come equipped with a motive that makes us pay attention to signs that there might be infectious parasites about, and then to avoid contact. Disgust is a voice in our heads, the voice of our ancestors telling us to stay away from what might be bad for us.[8]

It seemed that PAT was right! But, for a scientist, a eureka moment, with its shaft of light that reveals a previously hidden pattern, isn't enough. A theory has to be able to be tested to destruction; it has to lead to predictions that can be falsified. Just as we'd published a paper setting out PAT,[9] the BBC came knocking, wanting to make a documentary about human instincts. They offered us the chance to carry out a web-based experiment to test the disgust reactions of the good viewers of Britain and others around the world.

So while I was being filmed offering the charismatic TV presenter Robert Winston sterilized parasitic worms for lunch, behind the scenes we were hurriedly pulling together an experiment. The idea was to offer people pairs of photos that were broadly similar, but one of the images would be morphed to enhance the infection content. So we took towels and stained them with food coloring. In one version the stain was blue, in the other the stain was brownish yellow—looking like bodily fluids. In another pair of images, we took photos of Duncan, the series producer, either looking normal or slightly pinked up with damp hair and a few pimples. Another pair of pictures showed an empty Underground train and the same carriage full of people. A few control stimuli were added: a football, a cat, and a revolting image of a mouth in which flies had laid eggs in the gums. We mixed all the photos up and then asked people to rate them for disgust on a sliding scale from 1 to 5. When the survey went live on the night of the broadcast, the response crashed the BBC's server. In the end over 160,000 people from 165 countries completed it. Our analysis gave unequivocal results; every disease-relevant image scored significantly higher than the less disease-relevant image, and this effect held for every region of the world.

One final test on the website involved a large picture of a toothbrush and a question: who would you *least* like to share a tooth-brush with: your partner, the weatherman, your boss, your best friend, or the postman? We predicted that the people who were least personally known would be most likely to carry diseases that had not already been caught, so the weatherman's brush would be the most aversive, and one's partner's toothbrush, the least. We got the order right, except for one detail. Sharing the post-man's brush was most aversive, the weatherman's came third. On further reflection this made sense, as people probably feel that they know their weatherman, since he is welcomed into the home every evening on TV, while they mostly don't know their roving postman.

We made other predictions from the initial hypothesis: one that women would be more disgust sensitive than men, because our female ancestors had a double burden, to protect themselves and their dependent children from infection. This prediction held up: women scored every stimulus as more disgusting on average than did the men. We also predicted that disgust would fall off in middle age, at about the time that reproductive activity starts to tail off. However, what we found was a steady decline from a high disgust sensitivity at age eighteen or so, to a low one in old age, with no sign of a change in the slope of the line in middle age for men or women.

Cutting Disgust at Its Joints

Our study, the first to experimentally demonstrate that cues of infection predicted the disgust response, led to a flurry of press interest and brought me PhD student Mícheál de Barra. Mícheál wanted to see if he could use PAT to improve on previous attempts to find regular patterns in the disgusting, such as those by psychologists Paul Rozin and Jonathan Haidt.[10] He decided that we needed to go back to square one, to use the sharp knives of both epidemiology and evolutionary biology to dissect disgust and look for its joints.

We asked ourselves what kinds of places and things it might have been adaptive to avoid, so as to not get invaded by parasites—the ravenous body snatchers. How should the brain carve up the world of the disgusting? How should it recognize what to avoid, and how should it orchestrate its response? In other words, is the world of the disgusting just a jumble, a ragbag of the nasty and the infectious, as it appears at first sight, or does the brain recognize a structure in it?

An infectious-disease textbook became the source of a long and varied list of situations in which people might risk contracting an infection. Highlights included being licked by a stray dog, seeing cockroaches, indulging in promiscuous sex, encountering poor genital hygiene, seeing an oozing lesion on a friend's foot, being in contact with someone who is unkempt, eating food that has gone off, riding on a filthy bus, and hearing about open defecation. (Of course, it would have been neater to have done this with an exhaustive list of the diseases that infected our ancient ancestors, but no such list exists.) We set up a website survey and advertised it on Facebook, getting more than 2,500 people to record how disgusting they found the sixty scenarios. The responses tended to cluster together, and they did so in ways that were subtly different from what we had anticipated.

We had thought that the shape of the disgust response might reflect the routes of entry of pathogens into the body: that there might be a different type of response to the parasitic organisms that we might breath in, that we ate, that entered through the genital tract, and that got in through the skin, because each of these would require a different kind of behavioral defense, with distinct kinds of mental machinery. But the picture that emerged wasn't quite like that. We identified six categories of stimuli: people of unusual appearance who showed signs of sickness or deformity; infected lesions and bodily fluids; people with signs of poor hygiene; having risky sex; unfamiliar and possibly infected foods, as well as certain animals and insects; and finally, things in the environment that had come into contact with infectious agents (*fomites* are surfaces that vector infection). The table summarizes these categories.

While it wasn't exactly what we were expecting, this pattern still made sense. If you are an animal that cannot directly see infectious agents, then you need to be able to reliably avoid the places and things that tend to harbor pathogens. Those animals in our evolutionary past that did so would have passed on more

Table 1. Categories of disgust elicitors emerging from factor analysis of sixty stimuli

Type of disgust	Elicitors	Function
Atypical appearance	Abnormal body shape, sickness and sickness behavior, context (e.g., poverty)	Prevent interaction with infectious people
Lesions	Pus, scabs, sores, boils, wounds	Prevent interaction with infectious bodily substances
Hygiene	People's wastes, bad manners	Prevent uptake of pathogens from the environment; punish unhygienic behavior
Sex	Promiscuous or unusual sexual behavior	Prevent sexually transmitted diseases; favor optimal mate choice
Food/animal	Food or animals with infection or spoilage cues	Prevent consumption of, or interaction with, disease and disease vectors
Fomite	Things coming into contact with a disgusting object	Track contamination with potentially infective substances

Source: M. de Barra, "Attraction and Aversion: Pathogen Avoidance Strategies in the UK and Bangladesh" (PhD diss., London School of Hygiene and Tropical Medicine, 2011).

genes, on average, than those who did not. Since other people are the prime source of human pathogens, avoiding people whose appearance or behavior hints that they might have a disease is paramount. Any bodily substances that they exude should also be dubious, especially if these look or smell infective. It's particularly important to avoid the sorts of contacts with other people that might give you a sexually transmitted disease (as well as avoiding unsuitable mates). Animals and insects that vector disease, as well as foods that are unfamiliar or that show signs of being rotten, are best avoided because they are likely to carry pathogens. The data suggested that we also have a kind of disgust that is reserved for objects that have come into contact with things that are contaminated. Finally, people who behave unhygienically—who appear unkempt and poorly groomed or who cause contamination of the shared environment—are also the object of a separate kind of disgust.[11]

These six types of disgust were distinct; we think that each reflects a slightly different way in which evolution has organized the brain's response to recurring patterns of disease threats in the environment.

When we started this research, there were few people who agreed that it made sense to see disgust as an evolved adaptive system, and I had to battle my way through barrages of critiques in lectures and teaching. People would tell me lovely stories about how their babies would explore their own poo, arguing that this disproved the idea that disgust was innate in humans. Yet to take a comparable example, babies have little in the way of sex drive; it appears when needed, in adolescence, with the pubertal hormone spurt. Yet few would deny that a sex drive is innate, since it appears despite attempts in many cultures to quash it (at least in females). Babies only need disgust when they are old enough to explore the world independently, and when their mothers use disgust expressions to help potty train them, in about their second year of life. Something that is innate does not necessarily have to manifest itself at birth.[12]

And what about disgust and food? How can human disgust be a part of our evolved natures when surely it is culture that determines what people eat? Why will the inhabitants of Iceland eat reindeer nose worms, the Chinese rotten eggs, the Swiss rotten milk, and the Ugandans grasshoppers? In fact, all foods, and especially those of animal origin, are potentially disgusting, but we make exceptions for what is familiar. In every society infants accept what their mothers feed them, and then they grow up with that particular food repertoire.[13] The culturally tried and tested can safely be regarded as fit to eat. But any novel foods are treated with suspicion, to be rejected, sniffed at, or tasted only in small bites. Disgust of food is the default, for good reason; the system works on the precautionary principle that it is better to be safe than sorry.[14] After all, if you don't eat what is offered, you may miss a meal, but if you do eat something suspect or infected, you may lose your life or make yourself seriously ill. Better stick to Mama's cooking!

And what about toxins? It is often claimed that the disgust system protects us from toxins as well as pathogens. Yet the only toxins that are generally found to be disgusting are those that are associated with signs of potential infection, such as hydrogen sulfide, which betrays bacterial or fungal decomposition processes. Bitter or sour flavors are distasteful, but are they actually disgusting? They certainly didn't make any appearance on our disgust lists.[15]

Explanations of Disgust

If PAT explains most human disgust and also cuts it at its joints so neatly, how come this evolutionary-epidemiological approach isn't standard wisdom? There are lots of reasons. Historically, it was thought that our aversions were simply products of culture. More recently, evolutionary explanations of behavior have been given short shrift by social scientists, and psychologists in particular, at least until now.[16] Studies of disgust have been dominated by a few famous psychologists, and the standard model has gone unchallenged.

One of the earliest writers on disgust was Charles Darwin. On his specimen-collecting visit to Tierra del Fuego on the *Beagle* in 1832, he collected observations of interesting human behavior alongside his other specimens: "a native touched with his fingers some cold preserved meat which I was eating at our bivouac, and plainly showed utter disgust at its softness; whilst I felt utter disgust at my food being touched by a naked savage, though his hands did not appear dirty."[17]

From the safety and comfort of Down House, Darwin sent out letters to his global collection of correspondents. They confirmed that the disgust face—turning up the nose, wrinkling the mouth, and making as if to spit, as well as the emission of an "eugh!" sound—was common to Greenlanders, Native Americans, Malays, and Australians. He even recorded the same face in his own infant son. Based on the etymology of the word *dégoût*, however, Darwin proposed that disgust's original purpose was the avoidance of things offensive to the taste.[18] This idea that disgust has oral origins has become orthodox in the field of disgust studies.[19] Yet if PAT is right, then its origin must predate humans (Darwin himself thought baboons and turkeys could show disgust), and it must serve to defend all of the portals of the body—the skin, the airways, the genitals—from infection, not just the mouth.

While Darwin emphasized the universality of the emotion, the philosopher Herbert Spencer was one of the earliest to focus on the differences between what different cultures found disgusting: "Here human flesh is abhorred, and there regarded as the greatest delicacy; in this country roots are allowed to putrefy before they are eaten, and in that the taint of decay produces disgust; the whale's blubber which one race devours with avidity will in another by its very odour produce nausea."[20]

Anthropologists from Spencer's time onward made the differing taboos and prohibitions of different cultures the stock of their trade. A famous book by the social anthropologist Mary Douglas entrenched this view of disgust as a product of culture. In *Purity and Danger: An Analysis of Concepts of Pollution and Taboo*, which appeared to great success in the 1960s, Douglas suggested that dirt was the mirror of culture. According to her, anomalous objects and events that do not fit the local cosmology have to be rejected, and classed as dirty or impure, as otherwise they would pose a threat to the social order. "Dirt then, is never a unique, isolated event. Where there is dirt there is a system. Dirt is the by-product of a systematic ordering and classification of matter, in so far as ordering involves rejecting inappropriate elements" (36).

The caste system in India was Douglas's paradigm case. Human wastes are anomalous materials, neither alive nor dead, both belonging to, but rejected by, the person. Hence wastes such as feces are dirty and polluting. Humans that deal in wastes become symbolically impure; these castes then have to be kept apart, for fear of social disorder. Douglas's approach is still influential in anthropological and cultural studies circles.[21]

Sigmund Freud also thought a lot about disgust. For him disgust was a mode of repressing desire. In an essay on sexuality dating from 1902, he proposed that disgust served to defend the brain against its own baser tendencies.[22] For example, disgust of the bodies of others, especially of relatives, reduces the temptation to give in to lust at every opportunity. Learning disgust of feces during toilet training suppresses any temptation to want to contact or eat them. Children keep their animal id in check by learning to be disgusted at their own baser urges.

In 1927 Aurel Kolnai wrote the first scholarly work devoted

entirely to the subject of disgust.[23] Using the phenomenological approach of self-examination that was current at the time in Austro-Germany, he provided a careful account of his experience of the emotion. For him, biological disgusts included excreta, secreta, dirt, animals, and insects—especially crawling ones—certain foods, the human body in too close contact, exaggerated fertility, disease, and deformity. His moral disgusts included excess and satiety, lies, deceit and corruption, moral weakness, and sentimentality. But Kolnai's explanation was in the Freudian tradition. Disgust serves a balancing, psychodynamic function, in promoting the avoidance of excess, surfeit, and a desire for death.

As a practicing psychiatrist in 1940s America, Andres Angyal puzzled over the things his patients found disgusting. He found a pattern of disgust objects being "base or mean, capable of permanently permeating anything they have contacted, leaving a lingering, unpleasant after-sensation, even after washing the hands."[24] For Angyal, disgust objects often had a sense of the uncanny—being cold and clammy and implying death. From anthropological accounts, he concluded that disgust is more universal than culturally specific, and that it operates independently of any knowledge about disease or microbes. However, the nearest he came to a coherent explanation of the patterns he described was that wastes belong outside and it would be a perversion of nature for them to be taken back in.

Angyal's analysis was a foundation for the work of Paul Rozin, who could be called the "father of disgust." His body of work, later with his collaborator Jonathan Haidt, is the most complete account and has become the gold standard in the literature over the past two decades.[25] To explain the varieties of disgust, they enlist a variety of explanations, using Angyal and Freud, as well as Douglas and Darwin.[26] In their account disgust has oral origins, arising variously from a distaste of toxins, an evolved aversion to pathogens, and a fear of eating animals. The concern about animals, they say, comes from a magical folk belief that we might become what we eat. Cultural dietary and food preparation rules serve to separate foods from their animal nature and help us not to think too hard about the process that brought them to our plate. Otherwise we would not be able to consume them.

Rozin and Haidt argue that the fear of oral incorporation of animal products has since spread beyond the mouth to the rest of the body. Hence, inappropriate sexual acts, poor hygiene, death, and violations of the ideal body "envelope" or exterior form (e.g., gore, deformity, obesity) are also found disgusting. They claim that this is because humans have to deal with the existential terror of knowing that they are animals and, as such, are condemned to die. Disgust develops to repress thoughts of mortality—a Freudian idea where one part of the brain represses another. Hence, certain sexual acts and envelope violations remind us of our animal nature, and poor hygiene places us below the level of humanity. The Rozin and Haidt standard model of disgust also has two further domains: interpersonal disgust, which they say serves to protect the body, the soul, and the social order; and moral disgust.

Outside of science, writers in the humanities have made rich and potent contributions to the literature on disgust. William Ian Miller has a terrific description of what he calls "life soup" in his wonderful book *The Anatomy of Disgust*: "Temperature, it seems, disgusts precisely in those ranges in which life teems, that is, from the dank of the fen to the mugginess of the jungle; this is the range in which sliminess exists, for slime ceases to be when frozen solid or when burnt to a crisp. The temperature must be sufficient to keep the old life soup bubbling, seething, wriggling and writhing but not so great as to kill it" (64). However, despite Miller's description being spot-on for the conditions that nurture pathogens, he resists a biological explanation—preferring

Freudian interpretation. He concludes that "ultimately the basis for all disgust is us—that we live and die and that the process is a messy one, emitting substances and odours that make us doubt ourselves and fear our neighbours."

Robert Rawdon Wilson, a Shakespeare scholar, fearlessly explores some of the darkest regions of disgust, both physical and sexual, and asks why it repulses, but at the same time fascinates. He proposes many positive uses for disgust, for example, in humor that ridicules immoral behavior and in marking the boundaries of the acceptable. Yet despite the breadth and depth of his exploration, Rawdon Wilson has to admit defeat. He labels disgust "the hydra" because he finds it too complex to explain.[27]

While these works are fascinating and illuminating, all of them propose convoluted explanations of disgust, and all give short shrift to biological explanations. An increasing number of scientists—including Daniel Fessler, Debra Lieberman, Josh Tybur, Mark Schaller, Richard Stevenson, and my group—are exploring disgust from a modern evolutionary perspective. For us, there is a simple and parsimonious solution to the puzzle of disgust: disgust systems evolved to defend animals from attack by parasites, the tiny, usually invisible, predators that attack by stealth and eat their hosts alive. It is a brain system that orchestrates behavior in the direction of pathogen avoidance—whether the pathogens are in the environment, in other animals, or, especially, in other humans. It prevents the entry of pathogens through multiple portals: the skin, the airways, the genitals, as well as the mouth. No magical folk beliefs, Freudian repression, or existential denial of death is needed to explain disgust.

That the human disgust system is a product of evolution does not, of course, mean that it does not vary from individual to individual or from culture to culture. Individuals inherit differing propensities to disgust, and individuals tune their disgust responses over their lifetime according to experience and local cultural rules (especially when it comes to food).[28] Nor does it mean that disgust did not take on an important role in moral behavior as humans inexorably evolved into an ultrasocial species. Disgust can appear to be a hydra, utterly confusing in all of its extraordinary and powerful manifestations. But there can surely be little doubt that the disgust system does have a single basic epidemiological function: it evolved to orchestrate the avoidance of pathogens and parasites.

CHAPTER TWO

INTO THE HOT ZONE

May I never lose you, oh, my generous host, oh, my universe. Just as the air you breathe, and the light you enjoy are for you, so you are for me.

Primo Levi, "Man's Friend"

To make an animal takes proteins, fats, starches, fluids, and micronutrients. These ingredients combine to form a tempting calorie- and nutrient-rich dish for other animals to feast on. We are all familiar with the food web: larger, stronger, faster predators eat smaller, weaker predators, which eat smaller, weaker ones, and so on down to the herbivores grazing on the autotrophic plants, bacteria, or algae that fuel the whole system.

But there is more than one way to make a meal of another animal. Rather than investing lots of energy in the hunt and chase, some animals have evolved a less dramatic strategy—parasitism. These animals climb on board, worm their way in, and stow away. They then feast on a smorgasbord of tissues and bodily fluids, not to mention taking advantage of the shelter, transport, and mating opportunities offered by their hapless host.

The parasitic way of life is a pretty good one and explains why parasites outnumber predators on the planet, both in terms of number of species and in total biomass.[1] Imagine, for a moment, one of those BBC wildlife series where we see life at night through an infrared lens. The shapes of warm animal bodies show up bright red against their cool nocturnal environment. Pink birds fly through a dark purple sky. Lizards glow yellow or orange. David Attenborough says breathlessly to the camera, "And look at the glowing patch left in the nest as the owl takes off on her nightly hunt!" Now, instead of looking at the world through a heat-detecting lens, switch to a parasite-detecting lens. What does the world look like? In fact, it looks much the same, but in place of the birds and the lizards are silhouettes of parasites. The animal bodies are bright red.

Parasites are everywhere, infesting skin, tissues, and guts; even the follicles of your eyelashes teem with microscopic worms. Every free-living animal is a seething mass of parasites. Our parasite-detecting lens reveals not just the fleas, lice, and ticks hiding in the pelt of the animal we are filming; it also shows the worms in its gut, the microbes in its flesh, and the millions of viruses that infest its every cell. Seen through this lens, all animals light up bright red—they are hot zones full of parasites.

Yet most animals do a good job of staying whole, of keeping their delicious bodies to themselves, of staying alive, with their parasites under control, at least for long enough to procreate. No one has yet been able to build a detection system that can scan for tiny bugs and invisibly small microbes hiding inside living organisms. But animals do have systems for detecting and avoiding parasites. And their parasite radar must be trained on particular parasites, those that are particularly risky to those particular animals. Mice need to avoid mouse nematodes, not fish nematodes. Rhinos have to avoid rhino viruses and not human influenza viruses. Every animal has to be able to detect the types of parasite that are specific to its kind. So a well-designed animal should have a parasite-detection system that is capable of detecting not just any parasite hot spot, but those that contain the most threatening varieties of parasite.

But if parasites can't be seen and they don't give off any radiation that can be detected on film, what's an animal to do? For example, if a lobster meets another lobster giving off an odd odor, then maybe it shouldn't share a den with it, as it might be infected with a lethal virus. Or if a killifish encounters another killifish with black lumps all over its body, then perhaps it should find another shoalmate. If a salamander is hungry, perhaps it shouldn't risk dining on another salamander of the same species, as it might ingest pathogens infectious to salamanders. And a reindeer should probably migrate regularly so as to avoid eating grass contaminated with parasites' cysts in the droppings of other reindeer.

All of the animals that are alive today have ancestors that were good at parasite detection and avoidance. Those that didn't have those abilities simply got eaten up and so ended their genetic history. Animals filter incoming sensory information—sight, touch, taste, and smell—use it to compute likely parasite risk, and then respond to that risk, just as if they really did come equipped with parasite-detecting lenses. This skill seems to be found in all animals, humans included. And we humans have given our parasite-detecting devices a name: "disgust." Though we may have invested it with special significance and a special name, the human parasite-detection-and-avoidance system doesn't differ much from that of other animals, and surely it must share common ancestry.[2] We humans have a few unusual abilities, built on top of our animal abilities; like our capacity to imagine parasites, and to learn from what we imagine, and our skill in the use of microscopes (real parasite-detecting lenses). But for the most part, we behave as most animals do. So if we want to understand human disgust-related behavior, we should turn to other animals.

Animals have four ways to avoid paying the dire fitness costs of being invaded by body snatchers. First, they can avoid close contact with animals of their own species, especially when they are sick, because this is where the best-adapted and most infectious parasites are likely to lurk. Second, they can avoid other species of animal that might vector parasites that can jump from species to species. Third, they should stay away from places and things that might be contaminated with parasites or their progeny. And finally, particularly enterprising animals can alter the world they live in, in such a way as to make it inhospitable to parasites.[3]

Task 1: Avoid Others, Especially if They Are Sick

While there are various reasons why animals of the same species might cuddle up to one another, intimacy is far from a great idea when viewed through parasite-detecting lenses. Female house mice (*Mus musculus*), for example, take a good sniff of prospective mates, and if they detect a whiff of the protozoan worm-like parasite *Eimeria vermiformis,* they move on to the next male.[4] In one famous experiment, researchers painted red lumps on the wattles of the males of half of a flock of sage grouse (*Centrocercus urophasianus*) to mimic the effect of an ectoparasite infestation. These apparently lousy males had far less mating success than those that had not been so adorned.[5]

Choosing a healthy-looking bird as a mate has two advantages: it helps get good genes into your offspring, and it prevents you from catching something nasty, like a louse carrying a virus, in the here and now. An unhealthy partner could make you sterile or, worse yet, could introduce congenital disease into your breeding line. Disease avoidance therefore offers a good additional evolutionary explanation for why birds prefer healthy-looking birds as mates.[6]

Another way that animals test the health of a prospective mate is to provoke them to fight each other and see who comes out on top. Female squirrels and possums display their sexual availability prior to estrus, which leads to competition between males. The winner of the battle, which is likely to be the healthiest and least parasite-ridden, gets the girl.[7] Humans may be missing a penis bone for similar reasons. Having a big showy erection is a great way of displaying to a prospective mate that you don't have any fulminating diseases that could interfere with all those delicate hydraulics.[8]

In general it's advantageous to stay away from sick individuals of the same species. About 7 percent of bullfrog (*Rana catesbeiana*) tadpoles have a yeast infection that reduces their mobility and may lead to death. Given the choice, healthy tadpoles avoid going anywhere near those that have the infection.[9] Similarly, when experimenters injected killifish (*Fundulus diaphanus*) with black ink spots to mimic the effects of a common parasite, other killifish preferred not to shoal with them.[10] Caribbean spiny lobsters (*Panulirus argus*), usually social creatures, refuse to share dens with lobsters infected with the PaV1 virus.[11]

Parasite-detecting lenses are particularly helpful if you are a social species. While being social has its advantages—such as safety in numbers and the benefits of cooperation—it has a big downside in the form of greater risk of disease. Social primates are careful who they accept into the troupe; they will generally welcome a new member only after a long period of quarantine. During that time the troupe will often attack the outsider, testing its state of health. Any overt signs of sickness decrease the chances of acceptance.

Parasite pressure may actually place a limit on group size—in habitats rich in pathogens, such as the warm, humid rain forest, typical troupe size for colobus monkeys is about nine, while in the hot dry savannah of highland Ethiopia, with much lower pathogen loads, gelada (*Theropithecus gelada*) group size can run to several hundred.[12] Through parasite-detecting lenses, members of foreign troupes appear as parasite hot zones—especially because they might be carrying new pathogen variants, ones that the home group have no immunity to. Parasite pressure may be why primates are careful to limit contact with foreign groups, communicating only at a distance by calling, and by giving way to each other when they cross in the forest. Instinctive xenophobia may be a useful adaptation for a social species.

Another good way of not catching a parasite is to avoid meat, especially that from one's own species. Ecologists have puzzled over why cannibalism is so rare, observing that very few species satisfy their nutritional needs by nibbling on their neighbors. Parasites offer an explanation: one's cousin is a hot zone. A relative is more likely to carry an infection infectious to oneself than is a more distant species. The larvae of tiger salamanders (*Ambystoma tigrinum*), for example, have cannibal and noncannibal varieties, but the cannibals tend to carry much higher numbers of intestinal nematodes and bacteria than their noncannibal cousins.[13]

Humans, also, are adept at avoiding catching diseases from others of the same species. We sit as far as possible away from others at table or on trains; if someone shows any visible signs of disease, we tend to avoid contact and terminate interaction early;[14] and we turn cannibal only in extreme circumstances. Indeed, three of the six categories of human disgust response that our study identified concern others of the same species—people who look sick, abnormal, or disfigured; people as sex partners; and people who display poor hygiene.[15]

Task 2: Stay Away from Other Species, Especially Parasites, Parasite Hosts, and Vectors

Apart from their own kind, what further parasite hot spots might well-adapted animals avoid? Other animals that are also parasites themselves, that host pathogens, and that are used by pathogens as vectors all pose threats. Animals have evolved amazing repertoires of behaviors to defend against such risks.

Take *Caenorhabditis elegans*, for example. This tiny nematode worm, with only 302 neurons to its name, is much beloved by biologists as a model system for understanding animal physiology and behavior. This 1-mm-long creature is clever enough to detect a parasitic bacterium in its petri dish and turn around and flee from it, in seconds (see the film on the book website). When it is offered to nonparasitic bacteria to eat, however, it worms quickly over to gobble it up.[16] Ants are similarly discriminating, feeding on the corpses of other species, but scorning those infested with parasites.[17] Fish are known to avoid disease vectors; the rainbow trout (*Oncorhynchus mykiss*) can detect and swim away from parasitic eye flukes that cause blindness and, as a result, suffer fewer infections.[18]

The surface of an animal is like a tablecloth spread for a picnic, inviting hordes of hungry parasites to a free meal. Multiple species of lice, fleas, ticks, mites, bloodsucking flies, mosquitoes, leeches, bacteria, and fungi exploit or colonize the epidermis of every species of vertebrate. And vertebrates invest a lot of effort to get rid of them. Cattle stamp their feet and swing their heads in response to biting tsetse flies; fish scrape themselves on rocks and vegetation, as do elephants. Vampire bats (*Desmodus rotundus*) scratch to remove bat flies,[19] while birds preen, and impala (*Aepyceros melampus*) use their teeth as tick combs. When an experimenter stopped up the gaps in an impala's teeth on one side only, the side of the body that thus couldn't be groomed rapidly became tick-infested.[20]

Over 250 species of bird are known to "ant," rubbing crushed insects over their plumage. This distributes compounds that protect them from bacteria, fungi, and arthropods.[21] Gray squirrels and colobus, owl, and capuchin monkeys also rub their fur with leaves and fruit juices, probably for similar reasons.[22]

The effects of ectoparasite infestation can be serious: a cow calf (*Bos primigenius*) with a moderate tick load, for example, gains 10–44 kg less per year than a tick-free calf.[23] Bloodsucking mites reduce the body mass of house sparrow (*Passer domesticus*) chicks.[24] Apart from absorbing nutrients directly, biting insects serve as disease vectors; they introduce other, smaller, epiparasites such as the mosquito-borne plasmodium that causes malaria in perching birds and the tick-borne flavivirus that causes encephalitis in cattle. Parasites within parasites are a double burden, best avoided by all behavioral means possible.

Among primates, grooming to remove ectoparasites is of so much value that it can be exchanged for other resources like food or sex. Long-tailed macaques (*Macaca fascicularis*) have a biological market system where they pay for sex at the going rate in the currency of time spent grooming.[25] And what better time for a parasite to hop onto a new host than when the hosts are having sex? Ectoparasite transmission during mating has been documented in guppy, stickleback, sage grouse, pheasant, rock dove, barn swallow, grackle, zebra finch, and bowerbird,[26] not to mention humans.

Most animals need sex to produce offspring and can't avoid the disease risk that this brings. But they can take precautions. Primatologist Sean O'Hara observed that chimpanzees (*Pan troglodytes*) in the Budongo forest in Uganda regularly cleaned their penises, either with leaves or with their hands, after copulation.[27] And rats that are prevented from grooming themselves after sex catch more genital infections.[28]

Biologist Mark Pagel has proposed that being covered in a furry blanket—which requires constant grooming to keep lice, ticks, and other parasites under control—was so costly for some primates that, once they found other ways of keeping warm (fire, caves or clothes, for example), they pretty much gave up on hair altogether.[29]

In a hot-zone world, animals also face a real dilemma when it comes to deciding about food. On one hand, a morsel may be tasty and nutritious, but on the other hand, it may contain a hungry microparasite. This is one of the most ancient problems animals have had to solve, and, as with sex, each species has had to find a balance: a trade-off between the likely benefits and risks. To the oystercatcher (*Haematopus ostralegus*), the biggest cockles (*Cerastoderma edule*) are the most appetizing and easiest meal. However, the biggest cockles also harbor the most helminthic parasites, for which the birds are the definitive hosts. Ecologist Ken Norris showed that birds were feeding not on the smallest, least-parasitized cockles, because they were too much effort to open, nor on the fattest, but the middle-sized cockles, balancing the need for a cheap and a safe feed.[30] Butterfly fish (*Chaetodon multicinctus*), on the other hand, strike a different balance, actually preferring to feed on the bulbous lumps produced by coral infected with the cercariae of a tiny trematode. It seems that the extra energy gained from eating these fleshy extrusions that can't retract themselves like healthy coral outweighs the costs of ingesting more parasites.[31]

Predators have the same dilemma. It is much easier to kill and eat the sicker, weaker members of a prey troupe, but the predator that does so runs the risk of ingesting the parasite that made that individual sick and weak. Prey killed by predators are consistently infected with more trematodes, nematodes, and ectoparasites than randomly collected individuals.[32] Feasting on the sick and the dead requires investment in a robust immune system.

Humans, of course, also need to perform task 2. One of the categories of human disgust that our study identified is other species that might pose a parasite risk. We are repulsed by parasites themselves, when we can see them (including fictional versions in sci-fi horror movies); we avoid parasite hosts such as rats and parasite vectors such as cockroaches. And we are extremely careful about what we eat, especially when the food is another animal species or if it has been in contact with parasites.

Task 3: Stay Away from Parasite Hot Zones in the Environment

Conspecifics and other species are not the only places to encounter parasites. Animals that can detect and avoid hot zones in the environment have a comparative advantage in the race to get genes into the next generation.

Ants of the species *Temnothorax albipennis* avoid building nests in sites where they find dead ants, because corpses signal a possible hot zone.[33] If *Acromyrmex striatus* ants encounter a patch of fungal spores close to their nest, they close off the nearest entrance to help stop their nestmates from importing contamination.[34] The water flea *Daphnia magna* has to make a difficult and dangerous trade-off calculation: If it swims near the surface, it may be eaten by murderous predatory fish. If it swims near the bottom, it may encounter the spores of murderous bacteria lurking in the mud. In a neat experiment, daphnia were forced to swim nearer to the bottom of their tank by the addition of "extract of predator" to the top. They paid the price: picking up an increased load of microbial parasites.[35]

A nest is a handy adaptation for many species, providing shelter and protection from predators, but the downside is that your cozy home can also become a hot zone for parasites. Biologists in Switzerland offered great tits (*Parus major*) two kinds of used nest boxes to choose from. Half were infested with bloodsucking hen fleas (*Ceratophyllus gallinae*), while the other half had been microwaved. Of the twenty-three pairs of great tits that started breeding, three-quarters chose the parasite-free nests. The few that chose parasitized nests started their clutch an average of eleven days later, perhaps in the hope of outwaiting the fleas' breeding cycle.[36]

Nests can also harbor larger parasites—other birds. Cowbird chicks throw out resident chicks from a nest and assume their place; at least thirty-seven species of bird have been documented abandoning nests because of infestation with cowbirds.[37]

Environments that are contaminated with excreta are also likely parasite hot zones. Soils that have been fertilized with dung produce richer, lusher, more nutritious grass, but they also tend to contain more parasite larvae. Through parasite-detecting lenses, the greenest grass shines brightest. In tests, sheep avoided grass laced with feces that contained gastrointestinal nematodes. They became less picky about what they ate when they were hungry,[38] however, a phenomenon that has also been observed in humans. The parasitic potential of poo has even been evoked as an explanation for the phenomenon of animal migration. Reindeer and caribou may seek new pastures every year, not because of some mysterious wanderlust, but because they are looking for clean, dung-free pastures on which to feed, calve, and rear their young.[39] When I alluded to this explanation for migration at a dinner party hosted by anthropologists, one guest told me that Kalahari bushmen have similar worries. "It's getting dirty round here, time to move on!" she overheard one say to another. Another anthropologist related that Mongolian pastoralists do the same: timing migration to the buildup of human waste in camp.

Humans have one more, possibly unique (mice are another candidate),[40] means of detecting parasites in the environment. Humans pay attention when a parasite hot zone comes into contact with another object and remember what has happened. Like the infrared image of the hot spot left behind when a bird takes off, so humans remember the chain of contamination as if it were a series of hot spots. They can, for example avoid food that has fallen on the floor or a toothbrush that has been used by a stranger (we labeled this phenomenon "fomite" disgust in our web study).

Task 4: Modify the Environment to Discourage Parasites

There is one more strategy that animals can employ to reduce the dangers of parasitization—rather than avoiding them, they can make sure that hot spots don't arise. They can actively *modify* their environments to discourage parasites. An animal that has found a nice bit of habitat to feed and multiply in doesn't want it to fill up with wastes. Feces get in the way, contain toxins, and harbor parasites and pathogens. So what to do with your poo? As we've seen, you can just migrate and leave it all behind you. If you are a sedentary species, however, your ancestors will have evolved ways to deal with this problem.

Martha Weiss is the world expert on the poo-disposal practices of insects—although entomologists call it frass, not poo. She documents how leaf-cutting insects, like the caterpillars of the butterfly *Chrysoesthia sexguttella*, eat outward from the center of a leaf, leaving their droppings in the center, while those that eat inward, like the hispine beetle, leave a fringe of frass around the outside of the leaf. Her collection includes examples of "frass-flinging," "turd-hurling," and "butt-flicking." Skippers use hydrostatic pressure to fling their pellets up to thirty-eight times their body length away (153 cm for a 4-cm-long larva). Geometrid larvae hurl their turds with their thoracic legs, and noctuids jerk their abdomens to flick poo pellets twenty body lengths away. Butterfly larvae remove their frass by head-butting it away or by hauling it with their mandibles off the side of the leaf.[41]

Some animals are master compartmentalizers. Burrowing crickets use a specific corner of their chamber as a toilet and clean it up later.[42] Eastern tent moths (*Malacosoma americanum*) build silken latrines. They string huge webs across tree branches and use the lowest point as a toilet; when it becomes overloaded with feces, it detaches under its own weight and falls to the forest floor.[43] Fecal matter can also be put to good use. Some species of ambrosia beetle larvae pierce the walls of their cradles to eject feces, which the mother beetles carry off to manure fungus beds. Termites and some species of ants use their frass for nest building and for manuring their fungus gardens. Frass can even provide defense against predators. Cassidine beetle larvae exude a huge sticky wet fecal shield over their anal forks, to stop them from being bitten by predatory ants.[44]

Though fecal wastes are a nuisance to solitary or familial insects, the social species require sanitation systems. Ants and the other eusocial insects have to take parasite control seriously because they are both sedentary, having to live with their wastes, and highly related, making it easy for infections to spread. Most ants remove fecal material, as well as sick and dead colony members, from their nests.[45] The social crickets (*Anurogryllus muticus*) share a special latrine chamber, and social spider mites (*Schizotetranychus miscanthi*) always use the same spot within their nest for defecation.[46]

Eusocial insects are masters at engineering their niches to make them unsuitable for pathogens and parasites. The nests of most social insects have many separate chambers rather than one huge hall. Mathematical models show that dividing nests into a series of rooms helps slow epidemics of disease.[47] Ventilation systems help to the same end. Apart from modifying their physical environment to avoid the evils of excreta, insects can also modify their social environments—getting others to do their dirty work. Many species of ant have castes of cleaning workers who collect the feces, the sick, the dying, and the dead and carry them off to refuse piles a safe distance from the nest.[48] There are subdivisions of labor, with the ants that do the dirtiest work—on the midden—being segregated from those that collect the wastes. Any attempt by midden workers to socialize with others is met with aggression.[49] Older workers are more dispensable and are more likely to be found doing the dirty work.[50]

Sedentary fish, reptiles, birds, and mammals all have the same problems as insects—they need to engineer their environments to keep them from becoming parasite hot zones. They also build parasite-free homes, keep them clean, throw out wastes, and get others to help in the task—if at all possible. Some fish species invest energy in not fouling their living and eating areas. In the Red Sea, the surgeonfish (*Ctenochaetus striatus*) stops feeding every five to ten minutes and swims to a spot of deeper water beyond the reef edge to defecate.[51] Captive pike (*Esox lucius*) defecate away from the "home" area of their tank,[52] and damselfish (*Plectroglyphidodon lacrymatus*) defecate in specific sites around the edges of their small territories.[53]

Defecating around the edge of one's territory is common in many animals (for example, gecko, elk, and antelope) and is usually explained as scent marking.[54] However, it makes sense to keep parasite-ridden dung as far as possible from your feeding and living areas, as it does to deter rivals for your territory with the parasites that your dung may contain. Chimpanzees in zoos often throw feces at passersby, which may serve a similar function—to threaten rivals with parasites.

Parasites are a big problem for baby birds—nestlings are a juicy and defenseless feed for a variety of ectoparasites, including ticks, mites, and blowflies. Parent birds try to make sure that the nest is not a hot zone for parasites by defecating elsewhere and by removing nestling excrement, eggshells, foreign debris, ectoparasites, and dead nestlings.[55]

Most birds keep their nests clean of droppings. The chinstrap and Adélie penguins are a spectacular example. Like the frass-flinging insects, they stand up on the edge of their stony nests, turn their backs nest-outward, bend forward, lift their tails, and shoot out a projectile poo. The expelled material hits the ground about half a meter away from the bird.[56] In fact, emperor penguin (*Aptenodytes forsteri*) poo makes such a mess on the ice around nests that it can be seen from space, providing a useful means of monitoring the breeding success of this vulnerable species.[57] Barn swallows (*Hirundo rustica*) do it differently. Parent birds remove the fecal sacs of their nestlings and fly away with them, as can be seen on YouTube.[58]

Sometimes nests need more than just keeping clean: they need fumigating. Blue tits (*Parus caeruleus*) on the island of Corsica adorn their nests with fragments of aromatic plants such as lavender and thyme, which contain many of the same compounds used by humans to make aromatic house cleaners and herbal medicines.[59] These substances (linalool, camphor, limonene, eucalyptol, myrcene, terpinen-4-ol, pulegone, and piperitenone) have antibacterial, antiviral, fungicidal, insecticidal, and insect-repellent properties. Similarly, compounds in plants used for nest material have been shown to reduce the effects of fungi, bacteria, and ectoparasites on falcons and starling nestlings.[60]

Female great tits (*Parus major*) also spend a good deal of time sanitizing nests. In most cases, only the female birds do the nest cleaning—if a male great tit loses its mate, the nest soon becomes contaminated with remains of food, pieces of peeling skin, or even dead chicks, and the chicks are more likely to die.[61]

Some species even outsource their nest cleaning. One study found live blind snakes (*Leptotyphlops dulcis*) in 18 percent of the nests of eastern screech owls (*Megascops asio*). The snakes eat detritus and parasite larvae, which may make the owl broods healthier.[62]

Many species of animals thus modify their physical niches by cleaning up wastes. Some even modify their niches by influencing the behavior of others so as to reduce the threat of infection from parasites and pathogens—not unlike the cleaning and tidying behavior of the human animal.

From Disease Avoidance to Disgust

The animal world presents a stunning array of behaviors that help to prevent parasite invasion and infection. From selective feeding, to grooming, to frass flinging, to outsourcing cleaning to other species or castes, it seems that every animal that has been studied has "parasite-detecting lenses" and commensurate parasite-avoidance practices. While some of these behaviors could serve purposes other than avoiding infection, there is enough here to suggest that animals have a huge variety of infectious-disease-avoidance strategies. But is this "disgust"?

These behaviors are uncannily familiar, and the language used to describe animal disease avoidance is taken from the vocabulary of human behavior.[63] Some animals even respond in ways that look very like human expressions of disgust. Lab rats (*Rattus norvegicus*) gape, open their mouths, gag and retch, shake their heads, and wipe their chins on the floor when fed aversive tasting substances. Coyotes (*Canis latrans*) have been recorded retching, rolling on offensive food, and then kicking dirt over it. Monkeys react to offensive objects by sniffing and manipulation followed by breaking and squashing the item, dropping or flinging it away, and then wiping their hands.[64]

Given the overlap between what humans find disgusting and what animals avoid, and given that the same purpose is served (the avoidance of infection with parasites), should we keep the use of the word *disgust* for humans alone?

We are so used to thinking of disgust as a *feeling* that it seems odd to suggest that animals might have disgust, as we don't know if animals have feelings or not. But if disgust is reframed as the system in brains that drives parasite-avoidance behavior, in whatever species, then whether animals feel disgust or not becomes irrelevant.

If the function of disgust is the same across species, does this imply that the mechanisms that animals use to detect and avoid parasite hot zones are the same as in humans? Surely not. As with all adaptive features of all animals, some are similar because they share a common ancestry (homology), and some are similar due to parallel evolution (different solutions being found to the same problem). The systems that help ants avoid their infections will have little in common with the systems that make primates avoid theirs, for example.[65] Nevertheless, animals with which we share recent common ancestors, such as rats and primates, are likely to share some of the mechanisms that we use in implementing disgust. In the near future, when we better understand its brain mechanisms and its genetic determinants, it will be possible to construct a comparative phylogeny of disgust across the animal kingdom, showing what is shared and what is not, including *Homo sapiens* as but one branch on the tree. It is exciting that it is rapidly becoming possible to trace the deep ancestry of animal traits such as disgust.

While it is clear that disgust did not emerge fully formed in *Homo sapiens*, as many writers on the topic seem to propose,[66] we might still expect human disgust to have some special features. Humans alone have a much expanded prefrontal cortex, and we can use the imaginative ability that this gives us to apply disgust more widely (and perhaps more wisely) than can other animals. Humans are conscious of disgust; we have feelings about it; we are able to visualize and talk about it; we are able to learn from it and about it, and to make plans to avoid it; and we are able to weave it into our social and cultural fabric. In short, we humans are able to use disgust in new ways that are unimaginable to our insect, bird, mammal, and primate relatives.

CHAPTER THREE

DISGUST'S DIVERSITY

I believe that when man evolves a civilization higher than the mechanized but still primitive one he has now, the eating of human flesh will be sanctioned. For then man will have thrown off all of his superstitions and irrational taboos.

Diego Rivera, *My Art, My Life*

Dominating one wall of the Mollien gallery in the Louvre is an impressive oil painting by the French romantic artist Théodore Géricault. Adrift at sea on a disintegrating raft are fifteen figures, some hopeful of rescue, others in despair, others—gray-hued with limbs trailing in the water—already dead. A closer examination of the bottom left of Géricault's heroic composition reveals the torso of a man, with bearded head, shoulders, and arms, but missing his trunk and legs. Visitors to the Paris Salon of 1819, where the painting was first displayed to huge controversy, recognized this imagery immediately. They had all read newspaper accounts of the French navy frigate *Medusa* that had run aground off the coast of Africa. With too few lifeboats, 150 people were cast adrift on a makeshift raft. Those who survived endured dehydration, starvation, and madness, and some had resorted to cannibalism.

The *Wreck of the Medusa* was exhibited in London and drew huge crowds: some perhaps devotees of French romantic art, but most surely lured by the idea of contemporary French cannibals.

The consumption of human flesh is a nasty subject, hard to write about, even for a seasoned disgust researcher. But why this visceral response? If I had been raised in a society where cannibalism was normal, perhaps I wouldn't be averse to a tasty meal of "long pig." The apparent cultural specificity of cannibalism seems to challenge one of the main tenets of this book. Surely the notion of eating the flesh of another person should be disgusting to everyone, always and everywhere. And this should be because of our ancient evolved motives; animals that had propensities to eat their cousins also ran the risk of consuming their cousin's parasites. The more closely related the species you treat as meat, the more likely you are to catch something infective to you.[1] The idea of the edible dead should thus be abhorrent to all humans everywhere.

Yet this is a mistaken view of what evolution teaches us about human behavior. Yes, disgust may be a universal, a system in brains that evolved to help us avoid the fitness costs levied by parasites, but this does not mean that disgust works in the same way the world over, from culture to culture, from social group to social group, and from person to person. Though disgust has a main theme, it also has variations. Disgust sensitivity is an aspect of personality, varying between individuals. It fluctuates over the life course, and even by time of day. The specific disgust responses of individuals are influenced by individual experience; those who have had a disgusting encounter with a dead rat may respond more aversively to rats. Those who have had a bout of sickness after eating rotten chicken liver pâté may never want to eat it again.

Further, disgust is only one among many motives in the brain; it can be down-regulated when there are more pressing needs to attend to than disease avoidance. And disgust is not just a matter for the individual; we have hive brains where what others think and do affects what we think and do, and what we think and do about the disgusting, in turn, affects our culture.

While all humans are all equipped with a disgust system, the outputs of the system are not fixed, but flexible, within certain limits. Disgust is thus more than "just" an emotion; it is an adaptive system, natural selection's elegant solution to the problem of avoiding invisible infection in a range of different contexts.

Disgust Personalities

We have all encountered individuals who are too fastidious to drink from a shared water fountain, just as we have met others who care so little for their own hygiene that they cause offense with their poor manners and their bodily emanations. These personality variations emerged even on the raft of the *Medusa*. In their account of the disaster, Savigny and Corréard described how many of the raft's unlucky passengers could not bring themselves to consume the strips of flesh that had been pegged to the rigging to dry.[2] Those who could, however, found that it restored their energies, helping them to stay alive until they were finally rescued.

Why did some individuals eat human meat, while others found it too disgusting to contemplate, even at such enormous cost? Like height, disgust is a trait that varies between certain limits. Just as it was disadvantageous to have many of the genes that led to extremely short stature or to terrific height, so it was with disgust. Those ancients who were low on disgust sensitivity, who didn't worry about eating bad meat or about their bodily hygiene, were weeded out of the ancestral gene pool by disease. And those obsessives who couldn't tolerate even the slightest contamination of their water or who were too squeamish to touch any body fluids wouldn't have contributed many of their genes to today's pool either. They might have died of thirst and would certainly have found the intimacy of reproduction a challenge—let alone coping with changing their offspring's dirty diapers.

People differ from one other because of phenotypic and genetic diversity in a population. Individuals have ancestors with different numbers and combinations of the genes that contribute to the expression of disgust sensitivity, who lived in varied climes and settings with different levels and kinds of disease threat. Everyone therefore inherits a shuffled deck of genes for particular disgust dispositions, and this is a relatively fixed aspect of our personality.

Like other aspects of our personality, diversity can lead to difficulties. Discrepancies in disgust sensitivity are a major source of marital disharmony—for one partner, sharing a fork is unacceptable, while the other can't see what all the fuss is about. For one partner showering every day is essential, while the other is content to cultivate interesting odors. No amount of logical argument can persuade one of the other's point of view. Disgust sensitivity may, in fact, be a component of neuroticism—which is one of the famous "Big Five" dimensions of personality.[3] Neurotics tend to succumb more often to anxiety and depression,[4] so much so that personality researchers wonder why such an apparently maladaptive trait should persist today. Disgust may offer part of the solution to this puzzle. In our ancestral past, high levels of neuroticism might have led to more infection avoidance behavior, just as it would also have helped people avoid other dangers such as predators and accidents.[5] Though we don't know whether neurotic people are, in fact, better protected from infectious disease, something akin to this phenomenon has been observed in sunfish, *Lepomis gibbosus*. (Yes, fish do have "personalities."[6]) It turns out that shy sunfish carry different types of parasite than bold sunfish, presumably because their behavior affects the type of parasite that they encounter.[7]

History does not relate whether the lone woman on the raft, a sutler from the south of France, was high on the neurotic scale and was therefore unable to eat human flesh (though we do know that she was in a poor way and was thrown overboard before rescue). She probably had a stronger disgust response than the male castaways (remember that in our BBC web survey women scored disgusting images higher than men on average).[8] This could be a fundamental gender difference that reflects women's differing evolutionary history of responsibility for child care.

Disgust sensitivity may be a trait that is established at conception, but it can still vary over the course of a lifetime. Paul Rozin has documented how children start life with a low sensitivity to disgust,[9] only later developing a full-blown disgust response. This may be because disgust is not needed in the first few years of life, when an infant's exposure to pathogens in the environment tends to be controlled by its mother. Disgust sensitivity rises to a high in young adulthood and then tapers down with age. Teenagers constitute a huge market for hygiene products such as shampoos and deodorants, many of which are marketed as mating aids.[10] Again, this peak is probably adaptive, since it is helpful to be sensitive to disease cues during the mating and child-rearing years.

There are other times in the human life cycle when it pays to be especially avoidant of pathogens. Women in their first trimester of pregnancy lower their immune responses because they need to be able to tolerate the implanted egg, with its foreign genetic material. At the same time they become more squeamish about food, body products, and the risk of contagion.[11] It seems that there is a trade-off between the immune system and the disgust system; when one goes down, the other is ratcheted up. Indeed, the morning sickness of pregnancy may be a way of protecting both mother and fetus from toxins and infections during this vulnerable time.[12]

Dan Fessler and his colleague Diana Fleischman found that women tended to ruminate more about disease and contagion in the middle of the luteal phase, one week after ovulation—when immune responses are lowest—and ready to accept a possible pregnancy.[13] Cues to disease can also affect the immune system. When Mark Schaller and colleagues in Toronto exposed experimental subjects to pictures of either guns or infectious diseases, the white blood cells of those in the disease condition produced higher levels of interleukin-6, a molecule involved in immune defense.[14] The factor that links disgust with the immune system might be the neuropeptide serotonin, which is involved in learned aversions, vomiting, and the modulation of both innate and acquired immunity across Animalia.[15] Systems for avoiding and fighting disease may have very deep evolutionary origins indeed.

This cross talk between the disgust system and the immune system may even extend to what happens to an embryo during pregnancy—a subject known as fetal programming, now fashionable among health researchers. It seems that pregnant mice who are simply exposed to an infection, without even catching it, have sons who grow up to be less aggressive than the sons of mothers without the exposure.[16] Whether human mothers who have been in contact with disease during pregnancy give birth to less competitive sons, who thus save their energy to better fight off infection, is one of the many questions that remain to be explored by disgustologists.

Trading Off

"Disgustability" may be a trait that is part of human nature, but levels of disgust sensitivity vary with age and with the current state of the immune system. Do other factors influence how disgustable one is? For the survivors of the wreck of the *Medusa*, the biggest problem was to avoid death by starvation, which was a more pressing problem than the risk of catching an infection from corpse meat. Their disgust responses were thus probably partially suppressed: an adaptive "trade-off" that can be demonstrated in the lab. While ethical review boards don't permit one to starve one's subjects or to offer them "long pig" dinners, one researcher asked his subjects not to eat for fifteen hours and then shared pictures of nasty-seeming food ("looking like excrement or vomit"). He found that those who were hungry showed far less disgust, as measured by the activity of the lip-curl muscle (musculus levator labii), than those who had just had their lunch.[17]

What else might interfere with the disgust system? No one has explicitly examined whether feeling lusty reduces your disgust sensitivity, but a study of decision making and horniness by behavioral economist Dan Ariely suggests that it may. He asked male Berkeley students to imagine that they were sexually aroused and to predict their answers to a series of questions about what they would find attractive. He then asked them to masturbate, without climaxing, and answer the questions again. The students' predictions and their actual responses did not match up at all. While only 13 percent thought they would find sex with an extremely fat woman attractive in the cool state, the figure rose to 24 percent in the aroused state. Similarly, the idea of anal sex and sex with animals increased in attractiveness by 67 percent and 167 percent, respectively, in the excited state. Twenty-eight percent more students liked the idea of seeing an attractive woman urinate when they were aroused.[18] This ability of lust to down-regulate disgust was essential to our ancestors' becoming our ancestors—given how messy the business of sexual reproduction can be.

The Béarnaise Sauce Effect

The disgust system is a product of evolution, shaped by the hostile environments in which our ancestors evolved. The types of stimuli that were reliably associated with infection—for example, the smell of rotting human flesh, the taste of feces, or the sight of deformity—are innately labeled as disgusting. But parasites don't always betray themselves so readily. They may lurk in foods that smell fine or in people or animals that seem healthy, and they can spread surreptitiously over suitable surfaces. Faced with this invisible and unpredictable threat, how could the system learn what to avoid? If your disgust system got it wrong, it would mean that your genes would pay the highest cost: elimination from the gene pool. And if you happened to have genes for brains with psychological tricks that could help you infer the presence of dangerous parasites, you would tend to have more, and more successful, offspring, who would pass these genes on.

One of these tricks is known as the Garcia effect.[19] When he was a young psychologist working on radiation for the US Department of Defense, John Garcia noticed that rats that had been made sick by radiation were reluctant to drink the water in their cages. He experimented with a variety of foods and showed that the rats became aversive to anything they had consumed around the same time that they fell sick. He showed the same in garden slugs—which avoid carrot juice if it has previously been paired with lithium chloride.[20] This clever unconscious mechanism works for humans too—keeping us from returning to foods that have caused nausea in the past. However, as with the rats that mistook the food for the source of the sickness, not the radiation, the mechanism can misfire. The famous psychologist Martin Seligman recounted how he couldn't eat béarnaise sauce for ten years because of a bout of sickness following a steak meal. He *knew* that the sickness was probably due to the flu, as all of his colleagues had come down with it too. The intellectual knowledge didn't help: the associative conditioning had done its work, and every time he saw or thought of béarnaise sauce, it made him feel sick to his stomach.[21]

The Garcia, or béarnaise sauce, effect is a special case of associative learning—where an emotional memory is created about a particular stimulus. Imagine being offered a new and unfamiliar food on a foreign holiday—perhaps a slice of crocodile meat or an emu steak. Despite all your host's reassuring words, you'd be unlikely to tuck in with gusto until you had had a careful sniff and perhaps a small nibble. If you detected smells or tastes that you found disgusting, that would be that: polite thanks, but "no thanks." The primary reflexes of taste and smell decode the complex chemical signals emanating from organic materials that are possible foodstuffs, and hit the alarm signal when they find the signatures of biological toxins or pathogenic microbes. That alarm is recorded as a persistent emotional memory, and every time that same food is encountered, it's as if it has a flag stuck in it—"avoid this food!" On the raft of the *Medusa*, one of the castaways tried tasting excrement. The taste was so bad that he couldn't continue eating, despite his extreme hunger. According to Sigmund Freud, we all try poo tasting in infancy. But we only try it once; after that the warning flag in emotional memory makes sure that we never do it again.

It is not clear whether we come pre-prepared with the ability to discriminate disgusting from nondisgusting odors or not (it's hard to do such experiments on tiny babies).[22] Disgust responses to odors may be present in babies or may emerge later in life, when infants become more independent and need them more, or it may be that we start without them but are especially prepared to learn what is disgusting. Certainly it is hard to separate the two. If you search for videos of disgusted babies on the Internet, you find plenty of disgust reactions, suggesting innateness. However, if you listen to the sound tracks, you almost always hear the parent with the camera exclaiming, "Yuck! Yuck! Yuck!" at the smelly socks or dirty feet that the infants encounter. We seem to be highly motivated to teach what is disgusting to our offspring—another adaptive trait. The PAT of disgust predicts that we should have primary odor reflexes that emerge early, before two years of age, and that we should also be especially sensitive to learn what is disgusting from those around us and from experience.

Contagion

As far as we know, we are the only animals that can keep track of the history of disgusting objects.[23] Our parasite-detecting X-ray specs not only carefully note where disgusting things are now; they also record where those things have been in the past. So a cockroach that has walked across a sandwich leaves invisible footprints that make the sandwich inedible. It's hard to wear a stranger's clothes if we've been told that they haven't first been laundered. Many people feel that they have to wash their hands after using public transport, sensing contamination from surfaces that have previously been touched by multiple strangers. One of the creepiest stories in my disgust collection is that of a woman who bit into her hamburger in a famous burger restaurant chain, only to notice that a chunk had already been bitten out of the other side by some *unknown* other. Disgusting things like saliva, insects, or strangers leave invisible traces of contamination behind them, painting an indelible taint onto the object that has been contacted.

Perhaps the very reason why we experience the *feeling* of disgust is due to this need to learn what to avoid.[24] Each time we feel disgust, it's a bad experience that sticks a flag on the offending object, providing a reminder as to what to avoid next time. It's like a sticky emotional Post-it Note to the self saying, "Don't touch—bad stuff here." The associative flags can get attached quickly and easily, but removing them again is a slow process.[25] A therapist recounts the story of a man who noticed a discarded Band-Aid stained with blood in a hospital parking lot. Twenty years later he could still remember the exact spot where he saw it and was still avoiding it.[26]

David Tolin at Yale University demonstrated this "law of contagion" in a study using pencils and toilet bowls. He rubbed a pencil around a (clean) toilet bowl and then asked people to rate how aversive they found it. He then took another pencil, touched it to the first one and again asked how aversive it was. Taking the second pencil, he touched it to a third, and so on, up to pencil number 10. He found that for "normal" people the aversive effect fell off rapidly and had disappeared by the fourth or fifth pencil. For people suffering from OCD, however, the chain of contagion persisted; even the twelfth pencil was still contaminated. He describes the world of the OCD patient as being one of "looming waves of spreading contamination."[27] For the unfortunate sufferers, it seems that the law of contagion has gone into overdrive; even the slightest suggestion of taint spreads to every possible contact and hardly ever fades. Their perception of contagion can become so pervasive that it makes normal life impossible.

Culture: The Collective Brain

We all come equipped with high, medium, or low disgust settings, and different sensitivities to different kinds of disgust; life also tunes our system up or down according to how much we need it at the time, and according to our experience of encountering and learning about disgusting stuff. This is all very clever of our brains. But, smart as this is, we humans have an even better way of learning to be disgusted. One reason for our rampant success as a species is that we are more than just individuals; we are also wired into a collective brain. This hive brain is a fantastic repository of information, providing cultural maps that we use to navigate the dangers of hidden pathogens.

How does living in a collective brain affect our disgust sensibilities? And how far can culture influence those sensibilities? And to take us back to our castaways: Could the practice of cannibalism ever be so socially acceptable that we could all become cannibals? Could there even be cultures where the *Medusa* story would fail to elicit shock and disgust?

Individual brains affect the content of culture, and culture affects individual brains. While the collective culture of any one society reflects the psychological makeup of the individuals contained within it, it also reflects the shared experiences of those individuals and it reflects those ideas that people have found most profitable to pass on. More controversially, the collective culture of a group may also reflect a past history of being selected for as a group.[28] This can happen, at least in theory, when cultural ideas confer on groups the qualities that make them more successful and persistent than their competitor groups.[29]

Different groups have different cultures—as a result of their different experiences—and it is technically possible that there may be genetic differences in the underlying psychology of different groups. Imagine, for example, a population beset by famine. A small proportion of that population, by chance, were less disgusted by cannibalism. Those individuals survived the period of scarcity by eating the dead and dying. A population squeezed through such a genetic bottleneck would tend, on average, to have a lower disgust threshold and so be more predisposed to eating human flesh than neighbors who had not had the same experience. Whether that group would retain any distinctive genetic psychological differences would depend on the strength of the selection pressure and on the amount of population mixing after the famine event. A similar argument applies to disease. One society's constituent individuals could be more genetically disgust sensitive than another's because of a previous water- or louse-borne epidemic that led to the differential survival of the society's most fastidious individuals.

This is straightforward individual selection producing a change in the genetic makeup of the group. But now imagine a people who, for some reason, invented a mortuary rite that involved honoring the bodies of deceased elders by offering around small portions of their brains in a public ceremony. Imagine that this ceremony was shocking and powerful, and created a feeling of closeness among brethren, of social continuity, of family values, and of cohesion. Imagine then that such people had a competition over land with a neighboring tribe with no such ideology. Who would win, all else being equal? Clearly our cephalophagous group with its tighter social structure should have an advantage. The brain-tasting cultural ideology could then spread though conquest. This might help explain the persistence of the mortuary practices of the Fore people of Papua New Guinea, which continued for hundreds of years, despite being the cause of kuru—a degenerative brain disease contracted through the consumption of brain tissue.[30]

Now imagine a small sect that invents a ritual that involves pretending to eat the flesh and drink the blood of their leader. The rite is, again, emotionally "sticky," involving a collective experience of transgression, of overcoming disgust together, binding the celebrants into a tight brotherhood, with shared emotional memories. The brotherhood of the shared rite builds confidence and strengthens armies to conquer. The human tendency to follow charismatic leadership, and to blindly copy one another, does the rest. Conquest and evangelism spread the cannibalistic rite and the brotherhood across time and space, till Christianity becomes the dominant culture of the West and much of the rest of the world. (The substitution of bread and wine for real body bits removed the disease risk downside, of course.) There is some evidence that it was not just the ideas of Christianity that spread, but that, helped by such ideas, the genes of the Christians spread too.[31]

During the recent Liberian civil war, boy soldiers were reported roasting and eating human flesh. Acts of depravity serve to terrify enemies and to create a bond of brotherhood between members of a fighting band. Our own children can similarly indulge in *World of Warcraft*. The game currently has twelve million subscribers; players can choose to play as a race of zombies, called the Forsaken, which have a "spell" that allows the character to cannibalize corpses. In the equally popular *Grand Theft Auto*, media tycoon Donald Love has Toni Cipriani kill Avery Carrington and bring him to his private jet in order to eat him on the trip.[32]

All people, not just computer gamers, love to experiment with objects and ideas to learn how they work and what they do. Such "play" activity is generally accomplished in a safe environment, with a pretend object (a doll, a fairground ride), or in imagination, in a story (folktale, novel, movie, or computer game, for example). Emotional responses to simulated risks or experiences still get recorded as mental "Post-its" on objects or ideas, and they are then available to be called up when needed "for real." The profits to be made in the cannibalism industry—whether from books about Jeffrey Dahmer or Hannibal Lecter, from movies about zombies, from computer games, or indeed from the sales receipts for tickets to see Géricault's picture of the raft of the *Medusa* when it was exhibited in London—testify to the fact that "play" cannibalism is alive and well in Western culture.

But this evidence of cannibalism around us still hasn't proven that culture can make cannibalism an acceptable and normal part of social activity. We've seen examples of desperate cannibalism, deviant cannibalism, ritual cannibalism, and play cannibalism, but were there ever any societies where cannibalism was "normal"? Where cannibalism was practiced just for the sake of nutrition? Gastronomic cannibalism, as it were?

There are plenty of historical accounts, but it is very hard to find any solid evidence of normal, everyday, culturally sanctioned cannibalism. Though the historical literature is full of colorful tales of the cannibal tribes of Africa, the Indies, and the Caribbean (the Caribs gave their name to cannibalism), a closer look reveals how unreliable the accounts are. The typical report of cannibalism goes like this: "Our disgusting next-door neighbors are uncivilized savages, their women like nothing better than roasted baby, or boiled-up visitor, and they sleep around too." Or like this: "Tribe X were uncivilized savages who went around eating each other until we [insert name of colonial power] brought them order and civilization." In William Arens's investigation for his book *The Man-Eating Myth*, he found no reliable firsthand accounts of "everyday" cannibalism. He concluded that cannibal tales are just that—tales told by individuals keen to blacken the reputation of the despised enemy.[33] These are then faithfully recorded by the credulous anthropologist, wanting to believe the best of "his" pet tribe, and ill of its enemies.

Travelers also have a vested interest in being credulous, since shock-horror sells travelers' tales. Missionaries and colonists, too, have a motive for suspending their disbelief, since the worse the habits of the dreadful savages, the better the job they can claim to have done in bringing them to civilization. The flesh markets of eighteenth-century Fiji were probably a myth, and the Korowai of today's eastern Papua eat human flesh only in special circumstances. The archaeological record is similarly equivocal about cannibalism; cut marks on bones and heat-altered human remains have been found in many sites but can be interpreted in various ways.[34] The best evidence for the consumption of human flesh comes from the discovery of the human protein myoglobin in cooking pots and in feces in one hastily abandoned Pueblo Indian site in Colorado from AD 1150. But the fact that seven individuals were dismembered and eaten during, or soon after, the evacuation of the site again suggests drastic circumstances, not habitual flesh eating.[35] Cannibalism may only ever have been an act of deviance, a means to shock or to reinforce a ritual, or in desperation in times of famine—as in biblical Samaria or on the raft of the *Medusa*.

The ordinary, everyday consumption of other people for food has probably always been found disgusting, showing the limits to what can become a normal part of human culture. Cannibalism does, however, nicely illustrate the diversity of our disgust responses. They can be naturally strong or weak, they can be modified by experience, they can be suppressed—in extremis, they can be toyed with, and they can be elicited in powerful social rituals that shock and bind. Disgust is an adaptive system, standard equipment for humans (and all animals, as I have argued), whose basic themes around disease avoidance allow a degree of variation.[36] The disgust system is not infinitely elastic, however; we can never get rid of the ancestral voices in our heads that say, "Don't touch! Don't eat!" and call out alarms whenever infection cues, like the opportunity to consume human flesh, are detected.

CHAPTER FOUR   MANER MAYKS MAN

Manners are of more importance than laws. . . . Manners are what vex or soothe, corrupt or purify, exalt or debase, barbarize or refine us, by a constant, steady, uniform, insensible operation, like that of the air we breathe in.

Edmund Burke, *Letters on a Regicide Peace*

Imagine a world without manners. Getting up in the morning, your partner burps and drags on a grubby dressing gown. You can't find your toothbrush so you use his, and then wipe some muck off the bathroom floor with it. Downstairs, your daughter gives you a big dribbling kiss and then sits picking her nose at the breakfast table, wiping her fingers on the chair cushion. You select the least filthy plate from last night's pile and slop leftover curry onto it. Everyone grabs for the best bits, even before you can get it on the table, and the little one goes hungry again. On the way to work, you step over the large turd deposited by a neighbor on your front path and then drive into a jam caused by everyone ignoring the traffic lights. In your open-plan office, it's hard to concentrate for the flies and the smell coming from the rubbish pile in the corner, which has been used as a toilet by someone. You are distracted by a colleague recounting last night's sexual exploits over the phone in a loud voice. Your boss stands so close to you that you see the crud accumulated in his eyes and are overpowered by the smell of his unwashed armpits. Suddenly hungry, you reach over and grab your colleague's half-eaten sandwich, which you munch with your mouth open. In the conference room, everyone talks at once until a spitting match breaks out. Leaving work, the strangers in the elevator press up close to you, and one sneezes in your face. Home that evening, you slip in some grease that has been dripped on the kitchen floor; the kids have a farting competition while squabbling over the TV remote. Suddenly you can't bear it any longer, and it's time that everybody learned some manners!

Somehow all that is good about life is missing. Here there is no love, no caring, no respect, no community, no beauty, no quality of life. Without good manners, life hardly seems worth living. Yet manners are so ingrained in our lives that we hardly notice them.

Though there are plenty of self-help books on etiquette, explaining how to behave at a party, write a thank-you letter, or use table utensils in a foreign country, as well as tracts about how one *should* behave, such as Stephen L. Carter's book on civility,[1] few behavioral scientists have written about manners.[2] So where do manners come from, and what is their relationship with disgust?

One of the hottest topics in evolutionary biology is the question of what makes us human: what is it that makes us different from other animals? Some argue for language, others for tool use, others that fire allowed us to cook food to fuel big brains;[3] others suggest "theory of mind."[4] Undoubtedly, something important happened in the "great leap forward" about fifty thousand years ago that saw the change to truly modern humans with their ratchet of inventiveness, their cumulative shared culture, and their extraordinary ability to extract resources from the environment through cooperative enterprise.[5] For me, manners are a prime candidate for the key factor that distinguishes us from animals, one of the fundamental building blocks of what it takes to be a *Homo sapiens*. An old saw says, "Maner mayks Man." The acquisition of manners was the first baby step we humans took on the road to large-scale cooperation, and cooperation, underpinned by our moral sense, was the great leap forward that allowed humans to become a hypersocial species, a species that figured out how to work together and so to dominate our planet. So if we want to understand our modern world and the morality that makes it possible, we should start with manners, what the French call *la petite morale*.

Keep Your Distance

Looking at it from my perspective, when I meet you in person, I can respond in one of several adaptive ways. I can approach you and be friendly, signal that I mean well, and cooperate with you. On the other hand, it might be a lot better for me if I walked away, because you are a walking bag of microbes. With every exhalation, you might be emitting billions of influenza viruses or millions of plague bacilli; with every handshake, you might provide hundreds of salmonella bacteria or cholera bacilli; with your friendly embrace, you might be donating a dose of diphtheria or sharing scores of your scabies mites. Your immediate area might be full of your wastes, containing fungus from your fallen hairs or intestinal worms from your feces. If I wanted to get more intimate with you, I might contract hepatitis, syphilis, or worse. Proximity is potentially deadly.

Yet at the same time, you are a source of value to me. You may have objects to trade, gossip to exchange, social networks to tap into, or good genes to share. If I want to profit from associating with you, I have to take a measured risk—that these positive benefits will outweigh the negative costs of possibly catching something infectious. So I have to make a difficult trade-off calculation: shall I approach you or avoid you? And you too, of course, have to make the same calculation (albeit subconsciously), because to you I am a walking bag of human-adapted parasites that are looking for an opportunity to transmit themselves to you.

So how do we get around this problem? How do we maximize our chances of getting the most out of our social contacts, while limiting the risk of catching something nasty?

Let's try another thought experiment. Let's say you have invited an old friend to stay the night. What do you do to minimize disease transmission between you while keeping your relationship friendly?

The day your friend is due, you tidy the house, picking up unwashed underwear from floors and clearing away used tissues from the sofa, thus removing niches where microbial parasites might be lurking or breeding. You make a particular effort to scrub the bathroom and kitchen, where the sparkling tiles and shined taps display their freedom from bodily fluids. You vacuum up fallen skin cells, crumbs, and organic wastes from the carpet and put fresh linen on the bed, reducing the chance that she'll be bitten by one of your family's bedbugs or fleas. The white towel you place on her chair has been laundered clean of all traces of the family's bodily fluids, and a vase of lavender from the garden scents the air, masking the slight stale odor emanating from the microbial decay of the old mattress under the clean sheets. You brush your hair to signal that you groom carefully and have no ectoparasites, and pull on a dirt—and hence parasite—free blouse. You greet your friend with a warm hug, demonstrating that she is so valuable to you that you are prepared to run the risk of contracting one of her viral infections. With impeccable manners, she offers to take off her shoes, so as not to bring environmental pathogens into the house. With impeccable manners, in return, you say, "Don't bother," again to demonstrate that she is of more value to you than the state of your carpets.

You lay the table for dinner so that everyone sits as far apart as possible, to minimize the chances of sharing bodily emanations, and you give the wineglasses a special polish to make sure that there's no trace of saliva left on them. You are careful to serve potatoes to her with a clean fork, not the one that you've coated with microbes from your tongue. Though the dish is unfamiliar, your guest suppresses the desire to sniff at it, to check for fungal decay, showing that she is prepared to take a risk for the sake of your friendship. You remark that the coq au vin is a recipe that you learned from your mother, conveying covert reassurance that you know how to prepare food safely, according to the wisdom of your ancestors. Before dessert you carefully decant the cream from its tub with the out-of-date sell-by label into a pretty serving jug for the dessert. She chews her tart politely, with her mouth closed, so you don't have to catch any sprays of saliva or look into, and so think about, her microbe-filled cavities. She compliments you on your cooking and then jumps up to help clear the plates, politely offering to wash up. She helps you remove the saliva-contaminated food leftovers while you share the latest gossip over the kitchen sink.

Your guest asks to use the bathroom before bed and politely closes the door so as not to inflict the sound and odor of her excretions on you (because it would contaminate your brain with contemplations of the disgusting). She washes her hands with a new bar of soap you've put out for her, so that she doesn't have to use your family-contaminated bar or share her contamination with your family. As she wishes them good night, she keeps a polite distance from your partner and children, but gives you a quick hug.

The Dance of Manners

Your visitor is thus very polite, and so are you. You are each willing to accept some individual cost to ensure that you don't disgust or infect the other. At the same time, neither of you wants to appear more concerned with protecting yourself than with your friendship, so you demonstrate your intimacy by tolerating some risky contact. If you weren't both so skilled at the parasite avoidance manners dance, you wouldn't be able to be friends, and, as a result, you wouldn't get all the benefits that go with friendship: the warm interaction, the exchange of useful social information, and the mutual support in times of need.

This mannerly dance is played out every day, in every social interaction, in homes, schools, offices, factories, trains, restaurants, and shops. Yet we do the dance largely unaware of why we do it. We are not consciously aware that we are avoiding parasites, nor do we rationally calculate how best to avoid inflicting our pathogens on others. Instead, we have vague intuitions that it would be better not to disgust a guest by appearing unkempt, by showing them into a slovenly house, or by offering them a dirty towel. We know from previous experience that we would feel ashamed if we did such things and that we would feel particularly mortified if our hygienic lapses became the object of gossip. We try to behave in the "proper" way, and that usually means the way that we have been taught by our mothers and teachers.

The Origins of Manners

So if we are not aware of the disease-avoidance purposes of much of our mannerly behavior, how did we become a species that does what it does? At some point in our evolutionary past, manners did not exist. When did they arise? It is hard to be sure. Disease-avoidance behavior is ubiquitous in animals, as it certainly was in our ancestral lineage. Our mammalian predecessors discovered the benefits of social living and would therefore have had to deal with the problem of contact with conspecifics, who were prime sources of parasites. To avoid getting an infection, social mammals are known to avoid, or quarantine, the sick, the less healthy or odd-looking, and the poorly groomed.[6]

If animals can avoid potentially sick others, do they also "know" not to inflict their own emanations on others? While there certainly could have been an adaptive advantage for a social mammal in behaving in ways that reduce the disease threat that it posed to others, it's very hard to untangle behaviors that protect the self from behaviors that protect others. Did being well groomed, for example, help an individual baboon stay "in" with the troupe and hence provide a further explanation for the evolution of grooming behavior, beyond individual benefit? Certainly, early humans, like their animal ancestors, would have tended to avoid nasty, smelly individuals who failed to groom themselves or to keep their environments clean. Early humans were groupish, getting selective advantage from hunting or gathering in bands, for example. Those who were shunned by the group for their hygiene lapses would have paid a high cost in terms of exclusion from the benefits of social life. A society where the majority tended to avoid the dirty, ill-groomed, and slovenly would have set up a selection pressure on individuals to avoid being avoided. Those individuals who could better control their own emanations and thus appear less disgusting to others would have been better placed to profit from their social relations, thus gaining a selective advantage. This selection pressure, maintained over thousands of generations, would have led to changes in brains. Any tendencies that caused individuals to self-police their own hygiene—that is, to display good manners—would have helped them avoid paying the costs of social ostracism.

In humans this selection pressure probably spawned what we call shame. The shame system evolved because it predisposed us to learn to behave adaptively. Shame works by teaching us manners. Perhaps we went to school in a dirty shirt and the other kids called us names, or a guest refused to eat the food that we served that was past its sell-by date. The feeling of cringing shame was so painful that the offending behavior was never repeated.

Indeed, I suspect that one of the reasons that we are able to wear disgust on our faces and to make distinctive disgust noises is to communicate our disapproval of another's bad manners. It is to our advantage to get others to employ hygienic manners. By pulling a face and by exclaiming "*Yuck!*," we demonstrate disgust for another person's disease-threatening behavior. This elicits shame in the target, and they, as a result, modify their behavior.[7] Bad manners are not displayed, shame is avoided, and disease transmission is minimized. I suspect that we do this pretty much automatically, without needing to use our higher reasoning or, indeed, without needing to use higher facilities such as theory of mind.

Disgust demonstrations are especially important in the social training of our own kin, where the disgust face and the *yuck!* noise help us teach our children how to behave appropriately in public.[8] Every mother attempts to toilet train her infant, and most succeed by the time the child has reached the age of three—though it can be much earlier in some African societies.[9] This doesn't require giving lectures to toddlers about the dangers of germs in poo, but only a mother exhibiting a disgusted face and emitting a few *yuck!* noises when faced with inappropriate excretions.

But lack of hygiene is not just a matter for individuals and their families; it is also a threat to the wider social group. Conversations around the late Pleistocene campfire probably went something like this: "It stinks of poo around Og's house; I'm not going to visit him," or "Ig's kids are always filthy and are covered in sores; I'm not letting my kids play with them," or "Did you see the lice crawling about in Ag's hair? I'm not sitting next to her." And such gossip was of great interest, since humans are hypersensitive to information that is salient to their survival and reproduction. It's important to know who has the most hygienic behavior when you are looking for a play pal, a babysitter, or a mate for your daughter. Individuals in hunter-gatherer bands could thus acquire reputations as being clean or dirty types, and the latter would tend to be avoided or excluded, depending on the degree of threat that they posed to public health and on their willingness to reform their offending behavior.

This picture of Pleistocene tribes caring about the hygienic manners of others doesn't fit well with the usual stereotype of the ill-groomed, grubby, parasite-ridden, mannerless caveman. But all animals, including primates, are hygienic. Some of the very earliest human archaeological artifacts are hygiene-related, including combs, middens, and cave paintings of clean-shaven hunters.[10] Hygiene behavior is a universal attribute of all tribes in the anthropological record.[11] Sure, cavemen were dirty if you compare them with us today, but they probably did the best they could with no flushing toilets, power showers, shampoo, or steam laundries.

In a small-scale society, where everyone knew everyone else, there was a selective advantage to being able to avoid lapses of hygiene, since you would be welcome in society, rather than being punished by being excluded. The long juvenile period of the species *Homo sapiens* gave plenty of time for elders to beat hygiene rules into the youngsters, for their own good and for the good of everyone else in the group, and being equipped with the ability to feel shame meant that youngsters quickly learned to mind their manners.

Our own studies have found some evidence for a connection between punishment and poor manners. One of the categories of disgust that my PhD student isolated was for people who commit unhygienic acts, for example, failing to wash hands, to groom, or to defecate in a toilet. People who were especially disgusted by this category of infractions in others also scored highly on a psychological scale designed to uncover how punishment-oriented they were (for example, being more likely to want criminals to be locked up). We found a strong and significant association between the desire to punish and disgust for poor hygiene-related manners in others.[12] It seems likely that this connection must have originated in small-scale, reputation-based, Pleistocene society or even earlier.

The story of the evolution of manners saw animals being hygienic as a means of protecting themselves from infection, then early social hominids who stayed clean to avoid being avoided, and then modern humans who could gossip about hygiene and who felt shame if they upset others with their emanations. What happened next in the story of manners? As humans became more social, learning to live in groups larger than extended families, new problems presented themselves, and not just of disease.

Copy the Common

Let's try another thought experiment. Imagine that you are a new immigrant to a group; perhaps you are a new bride moving into your husband's family compound. How are you going to thrive in this new social setting? Proud as you may be of the skills that you've already acquired at home, it would probably be a good idea to figure out the local way of doing things. Maybe you should learn your mother-in-law's grilled fish recipe, pick up the local words for vegetables, copy the way your sisters-in-law toilet train their children, and note the correct way to address senior members of the clan. By copying behavior that is common, you are using strategies that have worked for others in this setting. By copying those who have high social status, you may also find success and curry their favor. If you fail to follow the local rules, however, you will remain an outsider, short on sympathy and, more importantly, on tangible support, when you need it. By adhering to local norms, you are not just more likely to be successful; you are also demonstrating that you have become an insider, "one of us," that is, someone who is prepared to invest time and effort in her new group, and hence unlikely to run off with the clan's valuables or its offspring. In short, you become someone who can be trusted.

This almost instinctual ability to mold ourselves to others, to scan for social norms and then follow them, has become part of the invisible glue that holds societies together. Given the multiple advantages that individuals gained from living in groups, those with strong copying tendencies (up to a point, of course) were likely to thrive and hence pass on more "copyish" genes to the next generation. Ig and Og around the campfire didn't have table manners, but they surely took care where they defecated and to wear their skins in the latest fashion. As settled agricultural societies became feudal and hierarchical, it became advantageous to curry favor with the elite, by copying their fads and fashions—what Elias has called the "civilising process."[13] As culture evolved, new ideas about hygiene and manners accrued; better ideas about child care and food handling, for example, would mostly have replaced ideas that were less beneficial.[14] And as technology advanced, it made more refined manners possible, fancy bathrooms and multiple types of eating implements led to conventions about how to use them, and industry learned how to exploit our copyishness to sell more jeans. Though we may not always be aware of it, we pay very close attention to the conventions of our social group. When we find ourselves in a new social setting—the tennis club, the parent-teacher meeting, or the holiday hotel, for example—we scan for the local mannerly conventions and rapidly try to adopt them.

Someone who wears a fresh robe to the temple does not just reduce the threat that he will spread his ectoparasites to the congregation, but the cut and color of the robe signals that he has paid attention to local tradition. Should he wear an outrageous color or a style worn by the tribe next door, he is signaling a desire to be different and will be less trusted as a result. Kids in most societies experiment with individuality as a means of distinguishing themselves but soon revert to following the mass, when they learn the lesson that only a few high-status individuals can get social benefits from being different. Wearing the "right" clothes to a wedding, meeting, or funeral is good manners, as conforming to convention allows group processes to proceed without distraction.

Small Courtesies

As group sizes grew from groups of related individuals, to clans, to whole tribes who came together from time to time for joint enterprises such as war or irrigation works,[15] the problem of cooperation with unrelated others became more serious. Individuals who tried to get the benefits of social life without paying their share of the costs could derail the whole cooperative enterprise. Humans became adept at looking for clues to who was likely to cooperate and who was not.[16] Manners provided an indicator. Those who were careful with hygiene and those who copied group norms were good candidates, as were those who went to lengths to show that they put the interests of others before themselves. The child who passes the plate of food before serving herself is not just sparing others from her pathogens and demonstrating that she is socialized into the rules of the in-group, but she is also showing that she puts others before herself. In effect she is saying: "Look how well my mother has taught me! If I can show such self-control now, how useful a member of this society will I be in the future! In the meantime you can safely do business with my family." And the child was no doubt taught by her mother that restraint with cake now would be likely to lead to her getting more total cooperative cake over her lifetime.

Such cooperative signals have to cost something to produce; otherwise anyone could fake them. By being continent with bodily fluids so as not to inflict our parasites on others, by investing time in learning the local conventions, and by making an effort to be courteous in small things, humans can demonstrate their ability to control their own greed, their readiness to put the needs of others before their own, and their willingness to invest in their social group. Youngsters who master these skills are set to reap all of the many benefits that come from living in a highly cooperative ultrasociety.

Manners are thus a signal of social intent. However, manners that concern hygiene have a particular force, as befits those that arose from the ancient and prehuman need to prevent disease in social groups. Philosopher Shaun Nichols hypothesized that norms around hygiene were likely to have greater cultural staying power than those that were simply local conventions, since the former are underpinned by the emotional resource of disgust.[17] He tested this hypothesis with a tract on manners, Erasmus's 1530 text *On Good Manners for Boys*. He classified these norms into two types. The first was about continence with bodily emanations. For example,

> "It is boorish to wipe one's nose on one's cap or clothing, and it is not much better to wipe it with one's hand, if you then smear the discharge on your clothing" (274).
> "Withdraw when you are going to vomit" (276).
> "Reswallowing spittle is uncouth as is the practice we observe in some people of spitting after every third word" (276).
> "To repress the need to urinate is injurious to health; but propriety requires it to be done in private" (277).

The second category included the following:

> "The person who opens his mouth wide in a rictus, with wrinkled cheeks and exposed teeth, is . . . impolite" (276).
> "When sitting down [at a banquet,] have both hands on the table, not clasped together, nor on the plate" (281).
> "If given a napkin, put it over either the left shoulder or the left forearm" (281).

Nichols found that the rules that concerned hygiene were statistically much more likely to still be in operation today, whereas the arbitrary conventions of the 1530s were not. So the rules of etiquette set out by Erasmus in 1530, with a disgust (and hence disease) relevance, are still recognizable today. Manners that were simply social conventions are more arbitrary, such as putting a napkin over the forearm or putting hands on the table at a meal, and have tended to disappear.

From Microbes to Manners to Morality

Without manners we are crude, rude, and uncivilized; we "gross people out," as one contemporary manners guide for teenagers suggests.[18] Without manners we threaten people with our unseen parasites; we drive them away from us and so deprive ourselves, and others, of the benefits of social life.

Manners have two fundamental components. The earliest (and the least recognized) content of manners is the behaviors that spare others the risk of catching parasites. As humans became more groupish, it also became advantageous to follow social conventions and to signal cooperative intent via small courtesies that implied willingness to bear the cost of putting the interests of others before one's own. Disgust and shame operate to prevent and to punish lapses in both of these kinds of manners.

Having an ability to learn manners about hygiene was primordial, the earliest problem humans had to solve on the route to hypersociality. Manners provide a basic framework of rules for sociality and so are a kind of mini-morality, a precursor to full-blown morality—the subject of the next chapter. That's why I place the dawn of manners as *the* key leap forward in human history, at least as important as the invention of fire or language. "Maners" do indeed "mayk man."[19]

CHAPTER FIVE

MORAL DISGUST

Cleanliness calls to cleanliness, clean houses demand clean clothes, clean bodies and, in consequence, clean morals.

C. E. Clerget, "Du nettoyage mécanique des voies publiques," quoted in G. Vigarello, *Concepts of Cleanliness*

Look at this list of what teenagers at my local school said was morally disgusting:

> Mocking someone in a crowd, bullying, racism, discrimination, standing by when others suffer, cloning babies, drug taking, suicide, bigotry, taking the Lord's name in vain, soldier deserting, spitting in someone's face, governments allowing injustice, putting profit before human life, exploitation of the poor, invading privacy, hypocrisy, back-stabbing a friend, poor sportsmanship, swearing, smoking when pregnant, vandalism, rape, adultery, treachery, stealing from the poor or handicapped, terrorism, chemical weapons, child slavery, cruelty to animals, pedophilia, incest, pornography, torture, wife beating, sadism, mistreatment of the elderly and vulnerable, cannibalism, eating dogs, murder, pollution.

The parasite avoidance theory (PAT) of disgust here seems to fail us. Why should this collection of immoral, hypocritical, cruel, exploitative, and nasty behavior be found to be disgusting? Were the kids just using disgust as a metaphor, or were they using their ancient parasite-avoidance systems in another context entirely?

There is a link between morality and disgust. To find that link, we have to start with the problem of human cooperation.

Adam Smith began his famous book, the *Wealth of Nations*, with the story of the pin maker: A single workman "could scarce, perhaps, with his utmost industry, make one pin in a day, and certainly could not make twenty. But in the way in which this business is now carried on . . . [o]ne man draws out the wire, another straights it, a third cuts it, a fourth points it, a fifth grinds it at the top for receiving the head; to make the head requires two or three distinct operations; to put it on, is a peculiar business, to whiten the pins is another; it is even a trade by itself to put them into the paper."[1]

Smith goes on to calculate that by sharing pin making among ten people, the men could, between them, make upward of forty-eight thousand pins in a day. Cooperation can thus render one individual 250 times more productive than if they worked alone.

Imagine, for a minute, a world of Smith's single workmen (and women), who do not know how to cooperate productively. In this world there would be no houses, just caves. There would be no beds, just piles of brushwood. Clothes would be missing; there would only be animal-skin wraps. No one would make breakfast for you; you'd have to dig something up for yourself with a stick and eat it raw. You'd have no transport, couldn't go to work; you wouldn't even have the language abilities that would let you speak to your family. You couldn't call the police, the exterminator, or the gas woman to fix the heating. You couldn't go shopping, take a vacation, or visit a hairdresser. You wouldn't be able to use roads, trains, or planes, and there'd be no phones, no computers, no art or music, no recipes or restaurants, and no pubs. In short, you'd be living as our primate ancestors did.

One way of measuring the difference between the level of advancement of our current way of life and that of our forebears is to compare energy throughput. A single hunter-gatherer uses energy at about the rate of 100 W, the bare minimum of subsistence food energy to fuel an individual living in a small band. However, the energy throughput of the average extant American is more than a hundred times this (12.7 kW). Our most advanced form of life is an astronaut; an inhabitant of the international space station uses energy at a rate five hundred times that of a hunter-gatherer.[2] To achieve such feats, we've harnessed the energy of plants, animals, water and wind, fossil hydrocarbons, and terrestrial as well as solar nuclear forces. None of this astonishing productivity could have been achieved by Adam Smith's pin maker working alone. Wherever you are reading this book, the lifestyle you lead and just about everything you see around you (unless you're a Luddite hermit on a pristine island), including this book, is a product of division of labor, of mass cooperation.

We are so used to our modern world that we hardly notice how miraculous it is. Whole populations of individuals make individual sacrifices to cooperate and divide their labor, so as to reap the later benefits. According to elementary evolutionary biology, this shouldn't be possible. In most species, an individual that helps another is just a fool, a nutter to be taken advantage of. Odd genes that throw up helpful tendencies that advance the interest of others over those of the self will simply go extinct. How has our species figured out how to suspend the selfish dictates of the genes and become supercooperative?

Simple selfless, helping behavior isn't too hard to evolve. There are lots of good evolutionary reasons why one individual member of a species should help another, even if this comes at a cost to that individual. If we help members of our family, we are helping those who have copies of the same genes, so we are helping our own genes.[3] This is the trick used by the eusocial insects, whose ability to cooperate on a large scale stems from the fact that they are closely related to one another. If we help others in the reasonable expectation that we will, in turn, be helped, then this helps our genes too. This is termed reciprocal altruism,[4] though the word *altruism* is something of a misnomer, as the individual interests are still selfish.[5] But why should we help unrelated strangers and people that may never reciprocate? A number of solutions to this puzzle have been proposed. It could be that we do such things to enhance our social reputations, hence giving us a chance to reap benefits from our kindly actions later on, so-called indirect reciprocity.[6] There is probably also a virtuous ratchet whereby prosocial people do well in a culture that favors helping, and the more prosocial people become, the more culture becomes prosocial. This is known as gene-culture coevolution theory.[7]

Critics worry that none of these explanations suffice to explain how we maintain such extraordinary levels of cooperation between unrelated individuals. Cooperation breaks down too easily; there is a strong incentive to cheat, to take the benefits of membership in the cooperative group without making your own contribution. Social order turns readily to anarchy. Game theorists have shown that one ingredient is essential to keep societies cooperating—you must make cheaters pay. Economists use public goods games, lab-based simulations of social dilemmas with real people and real rewards, to investigate cooperation. They find that with multiple players and repeated interactions, where people pay in but get more eventual payoffs if they cooperate, cooperation nevertheless gradually peters out, however beneficial it is. If, however, players are allowed to punish others who do not chip in their fair share, then cooperation ramps up rapidly and can be sustained over the long term. It helps if one individual punishes another for harming or failing to help a third person— truly altruistic behavior.[8] The evolution of punishment provides a mechanism for the maintenance of hypersociality.

Beyond kin selection, reciprocal altruism, and punishment, there is one other evolutionary process that may have played a part in helping humans scale the hills of selfishness and drop down into the Happy Valley of supercooperation. That is group selection. It's easy to see that when different human social groups compete with each other, those groups that are the most cooperative are likely to do best. The societies that can share labor to extract and process resources, that can invent and produce new productive or destructive technologies, and that can train and deploy effective armies will be able to outcompete the neighboring societies that are less cooperative. They will be more productive, better able to defend themselves from attack, and able to appropriate the lands, the goods, and the women of neighboring tribes. Their populations and the genes that they carry will multiply faster than the competition's. Cooperators will also be likely to find and like one another, shun the cheaters, be more successful, and so breed more cooperators. And if the most cooperative groups are relatively more productive, they will have more leisure time to devote to activities that help groups cohere, such as devising laws, operating institutions, and running religions. It requires quite special conditions of adaptive advantage and population mixing,[9] and is deeply controversial, but some sociobiologists are convinced that selection at the level of the group does account for some of our unique ability to be "super-cooperators."[10]

Whatever the exact evolutionary path we took to hypersociality, humans must have a mechanism in the mind that drives this cooperative behavior, which gives us our "hive brains." I'm going to call this the human moral system. Moral systems in brains evolved to make us *want* to be cooperators, to contribute to the group at our own cost, and to *want* to punish those who defect—so that we can ultimately reap all of those multiple benefits of being a member of a productive society. And because getting punished for our moral failings isn't to our own advantage, our moral motives also include shame and guilt—which provide preemptive internal punishment—helping us avoid being punished by others, not just for failings of hygiene manners, but for moral lapses too.

Gene-culture coevolution then further helped to domesticate us; our ancestors chose to mate with those individuals who were more cooperative and who chose to follow the rules of society, and the cooperators thrived better and had more offspring than the spiky anarchists. We have bred into ourselves qualities of docility, kindness, and sheep-like following of the local norms. It is as if we have antennae that look for and seek to learn the system of rules, of norms of behavior that are followed by others. We take some time to learn the moral rules, but, as with manners, we have a long childhood and adolescence in which to do it. We glean local norms from the evidence of behavior around us—learning from our parents, our friends, our teachers, and the written and unwritten rules of society. Moral systems reside not just in innate prosocial motives and learned behavior patterns in our heads, but also outside of them—in the heads of our parents and our role models, on stone tablets, in religious books and codes of law, and in novels, films, soap operas, newspapers, and tweets. Morality is what allows us to inhabit the superproductive Happy Valley of supercooperation. When morality fails, cooperation fails, and, like Adam, we are ejected from this Eden to fend for ourselves, to try to survive alone and unaided.

Moral Disgust: A Cheap Form of Punishment

If morality is the system in brains and in societies that evolved to underpin supercooperative behavior, then what is moral disgust for? How did it evolve and what adaptive purposes does it serve? To answer these questions, we have to return to the question of punishment.

Punishment, as we saw, is one of the keys to cooperation. Its job is to make it more expensive for someone to not cooperate than to do so. If I throw my old junk into the street, it has saved me the cost of disposing of it properly. If I am then fined or shunned by my neighbors, I pay a price that could well be greater than what I saved by not taking the rubbish to the dump in the first place.

There are lots of types of punishment: I could be insulted; I could be beaten up or locked up; I could be fined or even executed.

Active intervention, however, can be costly. My neighbor might hesitate to tackle me in the street about my rubbish habit because I could retaliate. If I was big and strong and had plenty of allies, I might give him a black eye, despite his efforts to uphold the values of virtue. In this case, what is the best strategy my neighbor can use to punish me and to return me to prosocial behavior? Two routes are left open. One is gossip—he can tell everyone that he meets about Val's rubbish habits and suggest that her behavior is shameful, a reflection of her bad character, implying that she wouldn't be great as an associate or trustworthy as a business partner. Contaminating my reputation would probably work well, at least in a stable community, as people tend to pay attention to such reports and use them to make judgments about whom to include in their social dealings. The second route is ostracism—Val can be avoided in the supermarket, and the residents' committee can debate her crimes and refuse to buy her jam at the street market. This is severe punishment indeed. In the film *We Need to Talk about Kevin*, the mother of the child sociopath is attacked in the street and has paint thrown over her house and car. But for Eva, played brilliantly by Tilda Swinton, the hardest thing to bear is her social exile; no one will talk to her at work, and the neighbors refuse to even acknowledge her existence.[11] By ostracizing someone, we can deny them all of the benefits of living in a social group and run little risk of bearing a cost for doing so, especially if we can "gang up" on that person through gossip.

When we encounter someone throwing their waste into the street or someone who has had a hand in a crime, we don't sit down and carefully calculate what cost to impose on the perpetrator. Hot anger and cold disgust well up inside us and tell us what to do. We make a rapid appraisal of the context and produce a suitable behavior: perhaps a loud insult if that person is small and defenseless, or crossing to the other side of the street to avoid meeting her, if not. We take every opportunity to discuss the crime and the criminal with others. Moral disgust wells up and tells us how to act; it tells us to punish the authors of anti-social acts by shunning them. It tells us to use the language of dirt and contamination to encourage others to feel disgust too. Disgust is perfectly suited to this job; it makes us want to stop, to drop the object of disgust, to steer well clear of it, and to recruit others to our point of view by insinuating that there's something nasty and contaminating around. Anger is the alternative, but anger can get you into trouble; your aggression can be met with retaliation. Disgust is cheap, and there's not much that the object of disgust can do to get back at you.

Disgust, then, is an emotion that first evolved to help us avoid parasites and then became an internal signal telling us to avoid and punish others who are sources of parasites—those with bad manners. From there it was but a short hop for the disgust mechanism to become a motive that makes us want to punish those who are behaving antisocially. Disgust expanded its object from microbes to manners and went on to become an essential component of morality.

Is Moral Disgust Really Disgust?

Though it makes a good story, this is still just a story. Is there any evidence that can help us determine whether the moral-disgust motive actually uses the same basic pathogen-disgust mechanisms and is not just a metaphor that people employ when they want to express disapproval?

It does seem as though there is at least some cross talk between moral and pathogen disgust. Psychologists Thalia Wheatley and Jonathan Haidt turned conjurer to hypnotize their experimental subjects into feeling a flash of disgust whenever they saw a particular word. When this word was incorporated into short vignettes about moral infractions, it made their judgments more severe.[12] Amazingly enough, the same effect was found in a later set of studies in the UK when subjects were asked to do the judgment task in a disgusting room or were seated next to a smelly waste basket, or when they had been exposed to a disgusting video.[13] Simone Schnall then went on to show the complementary effect, that people primed with a word task about cleanliness and purity made *less* severe judgments about moral infractions.[14] And in an intriguing study, Chen Zhong and Katie Liljenquist found that guilt could be reduced by hand washing, what they called the "Lady Macbeth effect."[15]

Some of these studies may be too good to be true, as it turns out that not all of their results can be replicated. A paper in the *Journal of Articles in Support of the Null Hypothesis* failed to reproduce Zhong and Liljenquist's results despite a huge sample size.[16] The jury is still out on this one.

Another possible source of evidence for a relationship between physical and moral disgust comes from the face. Exactly the same facial muscles—the levator labii—are employed in the lip curl of disgust at bad taste and images of feces, and at unfair offers in an economic experiment with a one-shot interaction called the ultimatum game. Is this really evidence that the two disgusts are one and the same thing, as the authors claim?[17] Again, we have to be a bit skeptical. We humans use facial expression as a communicative act—and the adaptive reason for communication is to manipulate others.[18] If you look up some of the wonderful videos of babies tasting lemons on the Internet (e.g., http://www .youtube.com/watch?v=Ixj88urcnIg&feature=related), you'll see expressively crinkled faces, charming wrinkled noses, and lots of tongues sticking out. The reason that the clips are so entertaining to watch is that the images fire our mirror neurons, making us cringe, smile, and laugh with the baby. In the babies' parents, they produce the effect that the expressions were designed for:

to manipulate them into stopping the lemon feeding (though sometimes only after several more video takes!). Of course, the disgust face also works the other way around; a parent pulling a disgust face can manipulate a child, for example, into discouraging it from taking a sweet from a potty.[19] And a disgust face directed straight at you tends to produce shame[20]—encouraging you to behave more morally.

But the most important, and sometimes insidious, result of the disgust face is the effect on third parties. We are much like the mice that notice and remember which mouse has been in contact with a diseased mouse, so as to avoid it later.[21] If I show disgust at what Adam is doing, Helen will pay attention, learn that what Adam is doing disgusts me, and remember that Adam did something disgusting. Helen will store the information to use when she next encounters Adam. The disgust face both manipulates the individual to make him stop the immoral action and manipulates others, encouraging them to treat the offender as morally contaminated and so shun him.

Providing evidence that the same facial expression is displayed in response to microbial and moral infractions is not evidence that moral and basic disgust is the same emotion. All that it tells us is that the same facial expression can usefully be employed in two different contexts. In an ultimatum game, a player may pull a disgust face, not because of some inner state, but because the face can influence other players—perhaps by making the perpetrator feel shame and so desist, or by implying moral contamination in the cheater—making the other players want to shun her. That same disgust face, as well as disgust sounds and words, worked to spread moral opprobrium when hunter-gatherers discussed Ig's dirty habits around the campfire, in the same way that it does now when office workers discuss the latest political scandal around the water cooler. "He did *what*? Ugh! How disgusting!" There is a connection between moral and microbe disgust—expressing it manipulates the behavior of others—but this still doesn't prove that they are one and the same thing.

Another way to figure out whether moral and microbe disgust really are the same emotion is to peer into our brains. Do the same brain areas light up in an fMRI scan when people are exposed to the viscerally and the morally disgusting? In one of the first studies to look at this issue, neuroscientist Jorge Moll found that that this was indeed the case.[22] Since then, neuroscientist Alan Sanfey has shown that the region of the anterior insula that is associated with disgust is also activated when subjects are made unfair offers in the ultimatum game.[23] The story is complicated because some of the regions are shared (anterior insula, basal ganglia) and some are different, as one might expect. And neuroscientists have tended to work from a motley selection of disgust stimuli, many of which are not pure disgust elicitors by our definition of disgust. Better studies comparing moral and microbe disgust are needed to clinch this one.

So, while we still can't be sure that moral and microbe disgust are the "same" emotion, there is clearly some overlap. The overlap is at three stages—with brain inputs, with brain processing, and with brain outputs. On the brain *input* side, many moral infractions involve elicitors of basic microbe disgust. Taking examples from the school list: rape, murder, torture, and cannibalism are all revolting crimes, and perhaps they are particularly revolting because they involve bodily fluids. In the brain *processing* box, there is certainly some overlap, in that people report that they are sickened and disgusted by moral infractions, and scans show similar brain-area involvement. And there is certainly overlap on the brain *output* side, with microbe and moral disgust eliciting similar use of words and facial expressions, and similar actions of avoidance/shunning and public pointing out of the contaminating nature of the disgusting object or person. On the principle that if it looks like a bear, sounds like a bear, and acts like a bear, then it probably is a bear, I'm prepared to take it, on current evidence, that moral disgust really is disgust, or a close relative.

It seems that the disgust system evolved first of all to help us avoid microbes, then its mechanisms were borrowed and extended to help us avoid and shun those people who posed a parasite risk through their poor manners, and then its role extended further to encourage us to avoid and shun those people who were behaving immorally. Shame evolved as disgust's counterpoint: helping us to behave in ways that stop us from disgusting others, so underpinning our manners and our morality. And then, because disgust focuses on contamination, it can be employed to label others as contaminated, making them morally contagious and, like the plague, to be avoided.

To find out more about moral disgust, we are running a web experiment called "the city of morals" with the BBC (you can participate at www.bbc.co.uk/labuk/experiments/morality/). An analysis of the first sixty-five thousand respondents shows that every one of thirty different morally dubious scenarios was scored as disgusting, as well as wrong. For example, someone setting fire to a museum, thus destroying priceless historical artifacts, scored an average of 8.8 out of 10 on wrongness and 7.7 on disgust. A person stealing a dress from a shop scored 8.0 for wrongness and 5.6 for disgust. When we ranked the results, however, there were some interesting patterns. Offenses that involved the spilling of bodily fluids—such as rape and incest, spreading flu, and failing to mend a broken sewer—came higher in the disgust ranking than in the wrongness ranking, as one might perhaps expect for parasite-relevant stimuli. But one other type of behavior also ranked more highly on disgust than wrongness. This category included a mother having children for the sake of welfare payments, a banker avoiding paying taxes, and a rich sports star not giving to charity. These offenses all come under the heading "social parasitism": people who blatantly exploit society for their own ends, without contributing fair shares. So perhaps there's another connection with the PAT of disgust here too. Moral disgust extends to parasites on the social system.

My current hypothesis is that all moral failings (i.e., defections from doing one's bit for society) occasion disgust, but that there are two categories of moral failings that have an extra-disgusting component: those that involve organic disgusts (acts of violence involving bodily fluids like rape, child abuse, murder, genocide, and torture) and those that involve social parasitism (like the ancient Greeks who begged bread at the temple when they did not need charity, who gave the name to parasites).[24]

Disgust is one of our moral emotions; it is a voice in our heads encouraging us to punish those who cheat, steal, spill bodily fluids, and exploit others without paying their fair share of the costs of sociality. As with other emotions, disgust evolved to bias our behavior, to help us make those hard choices about what to do next. Emotions lay bets as to what would be good for us to do now, based on the odds in the environments of our ancestors.[25] Antonio Damasio and others have shown that without emotions we are lost, unable to choose a course of action.[26] Disgust is one of these decision-making emotions, deeply entwined not just in microbe avoidance, but in our social life, through manners and morality. When we express repugnance at immoral acts, we are being deeply moral. We warn the potentially corrupt politician and the child abuser that the punishment of repulsion awaits, that if they persist they may become shamed exiles: a heavy, if not insupportable, price to pay for their selfish action.

Disgust is an essential component of human society. Yet our dual brains resist the idea that an irrational ancient emotional driver of behavior determines our modern morality. The philosophers argue that we are higher beings, able to make rational choices about right and wrong. The true story is that we are both rational *and* emotional. While immediate decision making is largely driven by the emotional brain, humans have also evolved new brain components that can see further into the future, imagine what would happen as the result of different courses of action, override emotional responses, and make choices that are sometimes, but not always, "better."

The disgust emotion, coupled with the mirror emotion of shame, remains a ubiquitous, potent, and effective deterrent to antisocial behavior. It is an ever-present, invisible threat to social defectors, and it underpins social order. Without moral disgust, we cannot dwell in our supercooperative Eden.

CHAPTER SIX

DISGUST MATTERS

This book has told a story. It has traced the history of the evolution of the powerful emotion of disgust from its origin as a defense against pathogens and parasites, to the role it plays in the manners that facilitate social interaction, to its expanded role as a moral emotion that holds together our complex and productive modern societies. But all good stories should have a moral. They should teach us a lesson, help us learn something that we can take away and use. So what is the moral of the disgust story?

For me, the moral of the story is that using science to unweave the complex rainbow of human behavior is a desperately important enterprise. Delving into the evolved drivers of human behavior offers many benefits: Understanding disgust's basic antimicrobial functions can help us to fight disease and to diagnose and treat certain psychiatric problems. Understanding the manners side of disgust can help to stop stigmatization of the sick and the different and so build more humane societies. And understanding moral disgust can help us to be fairer in our justice systems and perhaps in the end to be better human beings.[1] What could be a more important moral to this tale than that?

Using Disgust to Fight Disease

The most obvious benefit of understanding the disgust emotion is its practical use in the fight against infectious disease. While those of you living in developed economies have only a 5 percent chance of dying of an infection, if you live in Africa that chance is 65 percent, and in Asia it is 35 percent. This huge disparity remains an affront to modern societies and a challenge to public health practitioners, like me.

Diarrhea is one of the world's biggest killers, responsible for 0.85 million child deaths every year, mostly in the poorer regions of the poorer countries.[2] Perhaps because it is disgusting, and despite the fact that it kills more children than HIV/AIDS, malaria, and measles put together, diarrhea is still neglected. It's an orphan condition, with no global fund to tackle it, no film stars enlisted as global ambassadors to fight it, and insufficient research into how to defeat it.

Caused by a huge variety of pathogens, including viruses, bacteria, and gut parasites, the main source of diarrheal disease is the product of other people's guts. Human feces contain billions of microbes per gram. Hardly surprising, then, that poo comes near to the top of most people's disgust lists. These bugs have evolved to be past masters at getting out of one person and into another by catching a lift on every available sort of transport—hands, doorknobs, taps, water supplies, food, cleaning cloths, flies, cockroaches. As most mothers know, it's a losing battle to contain the runny stools of a child with diarrhea; the stuff gets everywhere, as it is designed to do by the microbes that it carries.

The best way of preventing diarrheal infection is to deal with the fecal peril. Feces have to be removed from human settlements, and any contact with feces needs to be followed up by careful hygiene measures. This may be easy to say, but, to the world's shame, fully 40 percent of the world has no safe place in which to defecate, having to go in the open or use unsafe or shared toilets.[3] And fewer than one in five people in the world wash their hands with soap after going to the toilet.[4] We calculated that if everyone in the world washed their hands with soap, it could save 600,000 lives a year.[5] On top of that, hand washing can help prevent respiratory infections, including SARS and pandemic flu, infectious blinding trachoma, infections associated with AIDS, and it can also help combat malnutrition.[6]

Of course, if you ask people whether they wash their hands, most people say yes, but in every study that we carried out, we found that actual rates were far lower than what people reported. Sitting in people's courtyards and watching what actually happens showed that only 3 percent of mothers in Ghana, 4 percent in Madagascar, 12–14 percent in China, Tanzania, and Uganda, and 18 percent in Kyrgyzstan were washing their hands with soap after using the toilet.[7] In the UK we saw that only 43 percent of mothers washed their hands with soap after changing a dirty diaper.[8] When faced with an interviewer, however, typically over 80 percent of people said that they did.

In studies in eleven countries including China, India, Vietnam, Senegal, Kenya, Uganda, and Peru, we explored the reasons why people washed their hands (or didn't) and found motives that included nurture, affiliation, and status. However, disgust at the idea that fecal material might be present on hands was consistently reported to be the most powerful motivator of hand washing with soap after using the toilet.[9] We took this idea to a commercial advertising agency in Ghana and used it to build a national campaign on hand washing with soap. We designed a TV commercial that showed an attractive Ghanaian woman exiting a toilet and then preparing and serving food to her kids. However, the film was doctored to add a purple smear to her hands as she left the toilet, which got transferred to the food that she fed to her children. We showed the first cut of the ad to mums in Accra, and the results were electric. We heard a sharp intake of breath and saw horror on their faces—feces were effectively being fed to children! Mothers were deeply shocked at this violation of the nurture motive and the evocation of disgust. After showing the ads on the three national TV channels for a year, a survey of the whole country showed that reported rates of hand washing with soap had gone up by 13 percent after using the toilet and by 41 percent before eating.[10] We found that most people knew about the ad, even if they didn't own a TV, and had discussed it with others. It had also encouraged teachers and health workers to take hand hygiene more seriously.[11]

Back in the UK, we wired up a public toilet in a service station on a highway to give us live data on how many people were using soap to wash their hands. We installed an electronic signboard that changed its message every hour, so we could see the effect that a series of carefully designed messages would have on the hand-washing habits of the quarter of a million people who came through during the study. The most effective of all the slogans in getting people to use soap was "Is the person next to you washing their hands with soap?"—attesting to the power of social approval and manners to influence behavior. Disgust-based messages such as "Soap it off or eat it later" also significantly increased soap-use rates, especially in the male loos.[12]

The UK government used disgust in their response to the swine flu epidemic in 2009/2010. They delivered to every house in the country a leaflet with an image on the cover of a man sneezing onto his hand, with a haze of droplets spraying from his mouth. The image was both attention getting and disgust inducing.[13] Telephone surveys showed that people who had been exposed to the campaign were more likely to buy antibacterial hand gel.[14] We continued to monitor hand washing in the highway service station during the pandemic; hand hygiene practices increased in line with media and blog coverage. However, they dropped back again to where they had started as soon as the social networks went quiet again about flu.[15]

Our recent hand-washing campaign in India uses a character called Laddu Lingam, a dirty fat man who makes sweets for children. In a skit acted out in schools and village meetings, he comes out of the toilet, doesn't wash his hands, and concocts disgusting sweets from dirt, worms, and dirty water. Not surprisingly, there are few takers when the sweets are handed around the audience. The campaign significantly increased hand-washing rates in the seven trial villages.[16]

Disgust has also been employed to great effect in campaigns to get people to build toilets in Asia and Africa. An approach known as Community-Led Total Sanitation begins with a health worker taking villagers on a tour of their village, and sticking small flags in the ground whenever they encounter a human turd. Some of these are collected and used in graphic demonstrations. For example, a hair is dipped into one, and then into a glass of water. The villagers are offered the water to drink, which they, of course, refuse. The point is made that open defecation is similar to adding feces to the water supply. This approach, which was developed originally in Bangladesh, sometimes pushes into more sensitive territory. In some villages, kids were given whistles to blow when they spot an open defecator,[17] which may be highly effective at instilling shame in the perpetrator, but presumably also leads to the social exclusion of the poorest in the village. Not a desirable outcome for a public health program.

Colleagues working in Cambodia designed a toilet-building campaign that used shocking images of people and dogs defecating, with the tag line: "A dog can't buy a latrine, but you can. Have a latrine yet?" And to bring more attention to the problem of toilets and diarrheal diseases, we helped organize a film competition called the Golden Poo Awards. The entries were full of imaginative uses of disgust.[18] *Time Magazine* commented that people might be offended by such efforts but asked which is more offensive: to talk about shit, or for millions of children to die because we *don't* talk about it?

While disgust's proper domain is that of infectious disease, it also works for other public health issues, such as smoking. The British Heart Foundation's most successful media campaign, entitled "Give Up Before You Clog Up," graphically depicted the impact of smoking on arteries by showing cigarettes dripping globs of fat.[19] The World Health Organization now recommends that images of diseased organs be carried on all cigarette packs. A Canadian study showed that the greater the disgust reported at such pictures, the greater the likelihood that people would have attempted to quit, and sometimes have succeeded.[20] Attempts to do the same in the United States were recently stymied because a judge ruled that this would violate the cigarette companies' freedom of speech.

Perhaps the most important reason that smoking has declined in developed economies, however, is that it has become bad manners. Once it became widely known that smoking harmed not just the smoker, but the bystander as well, smoking got added to the list of acts such as spitting and going ungroomed that are antisocial because they can cause disease. And people with bad manners meet with social exclusion. That exclusion for smokers is now even more explicit; they are ostracized, forced to visibly demonstrate their addiction to the world on office steps and in glass-walled pens in airports. It takes a lot of nicotine-fueled commitment to carry on smoking in the face of so much social disapproval.

Disgust can even be employed to combat obesity. In *Jamie's School Dinners*, a popular TV series on the UK's Channel 4, the celebrity chef Jamie Oliver, knowing that lectures about healthy eating weren't working, instead piled all of the snack and junk food brought to school by one class into one huge disgusting heap.[21] The campaign helped spur the UK government to improve the quality of school dinners. And in New York the consumption of sugar-rich soda drinks dropped by 12 percent after a public health campaign showing a fizzy drink turning into gobs of fat as it was drunk.[22]

Our approach to behavior change for public health stresses the power of emotions such as disgust, rather than rational drivers of disease avoidance. But emotions are powerful and need to be handled with care. Disgust may be a great candidate to employ in campaigns against disease, but it can offend or turn people off, rather than making them want to engage with a communication campaign. And it clearly has to be handled responsibly, since it can encourage stigmatization and exclusion of the individual involved in the unhealthy behavior.

Disgust's Downsides

While we're lucky to have vigilant disgust systems in our heads, ready to alert us to any sign of infection, as well as to be exploited in public health communication, the parasite-avoidance system doesn't always lead us to do what's best for ourselves or others. Disgust has plenty of downsides.

Like fear, disgust is an emotion with a hair trigger. Because it operates on the precautionary principle, it can lead us to reject things or people that it might have been better to interact with. Because it is so automatic and powerful, it can get in the way of carrying out essential tasks like caring for the sick or removing wastes. Because evolution designed it for a different era, it can misfire in our modern world. And because the system can go wrong, it can cause a variety of psychiatric problems. Disgust is a maker of misery, as well as of manners and morality.

The disgust system is tuned to overreact; it's better to miss one meal, or one mating opportunity, than run the risk of contracting a life-threatening disease.[23] Hence, disgust responses are precautionary and disproportionate to actual current risk. There have been running battles in California about the reuse of wastewater, largely because of scaremongering about contamination—the insinuation that almost infinite dilutions of contaminants might be harmful in water that has "passed through several people." Such arguments have derailed much-needed water purification schemes.[24]

As the food industry knows to its economic cost, even the slightest suspicion that products may have been contaminated leads to mass shunning that can affect a far wider range of products and goes on for much longer than is warranted by any actual health risk. Meat is already a suspect product because of its ability to carry and support the multiplication of bacterial contaminants; hence, when mad cow disease came along (which had nothing to do with bacteria), beef became suspect and consumption dropped dramatically, at a huge cost to farmers in the UK.

Indeed, disgust is probably the main reason that we throw out over a third of all of the food we produce. In a society where we are no longer hungry, the slightest sign of a mark on a fruit makes it suspect and hence unsalable in the supermarket, and the sniff test or the sell-by-date destines food for the wastebasket. I grew up in a society that made a virtue of thrifty cooking, so I'm shocked to find that some of my kids' friends won't eat anything that I make that contains leftovers. Babies, too, are deprived of good nutrition by overactive disgust. One reason for the decline of breast-feeding in the UK is that some mothers find the practice of feeding bodily fluids to an infant disgusting.[25] Breast-feeding is a lifesaver in low-income countries. But mothers we talked to in Zambia told us that they preferred to wean early, partly because they found the smell of breast milk spilt on clothes disgusting and feared that it would repel their husbands. Disgust is a strong affect; its sticky labels are easy to apply to things and very hard to peel off.

Doing the Dirty Jobs

Individuals whose work or lifestyle involve, or bring to mind, infectious fluids tend to suffer from sticky labeling too. The best-known example is the sweeper, or Dalit (previously known as the "untouchable"), in India. Studying the problem of sanitation in Lucknow, I covered my head with a *dupatta* and went out before dawn to watch the sweepers. As it began to get light, women and men could be seen moving from house to house sweeping steaming offerings from small holes in the backs of houses into baskets that they carried on their heads to empty into the nearest drain. No one noticed or greeted these shadow people; it was as if they did not exist. In Ghana, I learned that the same job of emptying bucket toilets could not be done by Ghanaians, but instead was done by members of a neighboring Burkinabé tribe, the Dagari. "Would you let your daughter marry a Dagari?" I asked Akan respondents. "No way" was the usual answer.

The toilets in my commuter train into London are tended by a range of non-Brits, taking a job that the natives don't want, as their first brave step on their way up the international career ladder. Being honest here, I'm not sure that I'd be happy to see my daughter marry a train-toilet cleaner either, for reasons that surely include contamination.

In France the word *putain*, which means "whore," is said to originate from the idea that prostitutes stink (*puer*) because of the sexual fluids that they absorb.[26] And females in general are suspect because they may be menstruating—which can spoil the sauce in Italy, stop bread rising in the north of England, and require monthly sequestration in Kenyan, Nepalese, and Native American cultures. The Old Testament forbids sex during menstruation, and a survey for a tampon manufacturer in the 1980s in the United States found that half of both women and men thought that one should not have sex during menstruation, presumably due to the double disgust of blood and sexual fluids.[27] A national survey in the United States in 1993 reported that 89 percent of adolescent males found the idea of sex between two men disgusting[28]—disgust of sex and bodily fluids spreads and sticks—contaminating the very gay identity. The negative effects don't stop there. Disgust's hypervigilance can be tripped by a range of conditions like epilepsy, mental illness, mental retardation, obesity, skin conditions such as psoriasis, cancer, and HIV, all hard, if not impossible, to catch via social contact.[29] Though it may have been adaptive in our ancient past to flee strangers with signs that they might be carrying an infection, such responses are wrongly calibrated for a modern world where infection is much rarer. Individuals with disabilities or disfigurements automatically activate thoughts of disease in onlookers, even when the perceivers know perfectly well that these individuals do not have infectious conditions.[30] People who are more concerned with disease are less likely to have friends with disabilities, to dislike obese individuals more, and to display implicit ageism.[31] Human faces made up to look sick are found to be more disgusting than healthy counterparts.[32] As a consequence those who are sick face a double burden—they suffer with the condition, and they suffer discrimination from those around them—or at least, they fear possible discrimination.

Shame, as the flip side of disgust, leads people to sequester themselves, to avoid disgusting others. Teenagers try to hide the acne that causes shame and poor self-image, and women with obstetric fistula (vaginal damage that causes incontinence, which is common in places with poor obstetric care) sometimes remove themselves from society for fear of causing offense.[33] Beyond the suffering of sickness, hospital patients often have to deal with the shame of public evacuation. One doctor recounted the problem from his own experience: "To lay in bed, and against all physical rules, and I may say psychological rules as well, and do what you normally do at the toilet was a humiliating experience of the helplessness patients feel when help with basic functions is needed. Why did I never question this part of caring when I worked as a doctor? For us, defecation was only an abstract category in the patient's medical record."[34]

Terminally ill people often fear losing control over their physical functions and hence becoming dirty and "untouchable."[35] So the old, the frail, the sick, and the disabled, all those who have to hand their body care to others, have to live in shame, fearing the disgust that they may cause.

Policy makers sometimes dismiss such fears as simply irrational, effectively asking people to "get over it." But this is to make a major error about human nature. Rules about continence with bodily fluids are based on powerful ancient emotional drivers of behavior, as well as on the manners that are inculcated into everyone at an early age. Fear of shame is a serious and a seriously debilitating matter that can't just be reasoned away.

And while patients may fear the disgust response of the caregiver, the caregiver also has to deal with her or his own disgust response. Caregivers have to overcome revulsion for the bodily fluids of those they care for. Sickness can strain relationships, for example, when the partners of AIDS patients find themselves having to deal with the messy processes of disease.[36] Despite their often heroic denials, it is clear that caregivers too suffer from stress and pay an emotional cost in overcoming disgust. To dismiss this as irrational is, again, unhelpful. Caregivers need the emotional labor of their work to be recognized, discussed, supported, and rewarded.[37]

Beyond those who care for the sick, professions that involve encounters with disgust elicitors also involve emotional labor. Sewage workers; launderers; road sweepers; toilet, office, and hospital cleaners; sex workers; emergency-service crews; abattoir workers; pest controllers; and morticians suffer both from stigma and the stress of overcoming disgust. They find little sympathy and social support, as few stalwart friends are ready to hold extended conversations with them about the horrors that their professions force them to encounter. Professional support tends, again, to concentrate on dismissing disgust as irrational, rather than engaging with its real, and often powerful, effects on the psyche.

The horror and shame of breaking rules of good manners affects young women who reach the age of menarche. When I hit that transition, I stayed in my room, refused to go to school, and claimed to be sick for several days each month. Though there are few studies on this, it may be a common pattern. Many girls in developing countries drop out of school at this time.[38] With little in the way of suitable toilet facilities and reliable sanitary napkins, girls live in fear of shame because toilet doors don't close (if there are toilets at all), makeshift menstrual pads leak and can't be washed and left out to dry in public, and boys tease. Faced with such problems, or even when there are menstrual-friendly modern facilities, girls the world over may prefer to sequester themselves in a sort of self-disgust when their periods start. Well-meaning NGO campaigns that seek to tell girls that they are behaving irrationally are again missing the point.[39]

When Disgust Goes Wrong

Like any of our organs, the disgust system and its components can malfunction. Genes, wiring, chemistry, physiology, and life experience can all conspire to make brains produce behavior that is maladaptive, interfering with the normal pursuits of everyday life. People at the top or bottom end of a normal distribution of disgust sensitivities might also find that they are classified as abnormal. Whether due to malfunction or finding themselves at an extreme end of a spectrum, many people have problems that relate to disgust. High levels of disgust make some people so squeamish that they can't leave the house, eat, or make social contact properly, and low levels of disgust make people unhygienic, unmannerly, and hard to be around. The fact that disgust seems to be made up of a number of subcomponents suggests that we might find pathologies associated with each: food/animal, sexual, lesions, hygiene, other people, and contamination disgust. Do malfunctions in each domain manifest as specific phobias or anxieties?

Obsessive-compulsive disorder (OCD) is an obvious candidate for a disorder of the contamination subcomponent of the disgust system.[40] OCD patients suffer from an excess of disgust. Thoughts of contamination and impurity intrude into their daily lives, and they try to reduce their distress by sanitizing and disinfecting themselves and their environment.[41] Sufferers describe how contamination is a constant concern—how they live in a world of spreading, looming contagion.[42] Not surprisingly, the condition gets worse in the presence of a disease threat—psychiatrists report increased distress and the exacerbation of washing rituals during epidemics, such as the swine flu alert.[43] And because OCD occurs along a continuum, it is likely that for every individual diagnosed, there are many more who suffer from some form of debilitating contamination anxiety.

As ultrasocial beings, humans depend on others for survival, yet other people, as disease sources, are a key subdomain of disgust. Social phobias may therefore be related to other-person disgust. Though most phobias go unreported, at any one time 4.5 percent of Americans may be suffering from social phobias and 2.3 percent from agoraphobia.[44] Abnormal unwillingness to venture into crowds and to contact other people is associated with heightened disgust sensitivity.[45]

Other specific phobias also relate to components of the disgust system. Blood-injection-injury phobia is characterized by extreme aversion to the sight of blood, injuries, or surgical procedures, including injections. Sufferers rate disgusting images as more disgusting than control individuals do, and they display stronger facial expressions of disgust.[46] Such sufferers would be likely to score particularly highly on our lesion subscale of disgust.

Animals that have connections with disease and dirt are much more likely candidates for phobias and childhood fears than animals that do not (i.e., spiders, rats, worms, maggots, cockroaches, teeming insects),[47] suggesting a disorder of the insect/animal disgust domain. Though there has been controversy as to whether spiders occasion fear or disgust, it now seems that disgust is a stronger predictor than anxiety of spider avoidance.[48]

Trichotillomania may also be disgust related; people with a condition that involves the compulsion to pull out skin hairs may be responding in an exaggerated manner to the possible presence of ectoparasites in skin—a hypothesis that has some support in the literature.[49] Another odd phobia has Internet groups dedicated to it. Trypophobia is fear of clusters of small holes. Looking at the images that trigger this phobia, I find I'm also scratching my skin and feeling slight nausea in sympathy with the phobics. To me the patterns of holes look like clusters of insect eggs, possibly laid in the skin. These patterns hyperstimulate my parasite detection mechanisms and creep into my nightmares. Look it up on the Internet if you dare check your own responses!

Since food is one of the subdomains of disgust, one might expect food anxiety disorders to relate to disgust. Clinicians point out that anorexia and bulimia feature disgust, and some, but not all, studies have shown heightened disgust in food phobics.[50] Meat is one of the most likely sources of pathogens in food and is also a special focus of anxiety. All cultures have taboos about what meats are suitable to eat, and vegetarians/vegans reject it entirely.[51]

Since sex is a subdomain of disgust, one might expect pathologies of the disgust system to affect sexual function. Though the problem has been little studied, several authors report that disgust is implicated in undermining sexual arousal and desire.[52] Clinicians recount case studies of women who had turned disgust on themselves, associating dirt, disease, fistula, and defecation problems with their vagina, leading to an inability to face intercourse.[53] The UK broadcaster Stephen Fry describes why he chose celibacy in the following terms: "I would be greatly in the debt of the man who could tell me what would ever be appealing about those damp, dark, foul-smelling and revoltingly tufted areas of the body that constitute the main dishes in the banquet of love. . . . Once under the influence of the drugs supplied by one's own body, there is no limit to the indignities, indecencies, and bestialities to which the most usually rational and graceful of us will sink."[54]

If the psychological problems that we have been talking about are pathologies of different components of the disgust system—of fomite, food, animal, sex, and other-people disgust—then we would expect some cross talk between conditions. A quarter of OCD patients in one study were found to be virgins, and 9 percent had not been sexually active for years.[55] Many also suffered from extreme shyness, suggesting possible social phobia comorbidity with OCD.

If these phobias can be explained as maladaptive overvigilance in the parasite-avoidance system, then are there also people with underresponsive disgust? It's been suggested that one of the reasons that patients with Huntington's disease can be hard to nurse is damage to the disgust system in the anterior insula.[56] Caring for such patients is challenging when they pay little attention to their toileting and keeping themselves clean.

Self-neglect is a common problem that social workers have to deal with, when clients' poor hygiene manners leave them isolated, ignored, and sometimes abused by society.[57] Indeed, I agreed to speak at a conference on this topic, but it was canceled—because too few people signed up to attend! Such disgusting topics are unpopular, unattractive, and neglected, as, indeed, are those who suffer. Individuals with learning difficulties have been shown to get a poor deal from the UK national health service. It is likely that people who have problems with personal hygiene—whether due to low disgust, autism, or some other issue—are likely to be discriminated against in all walks of life. This is one of the last bastions of social prejudice that needs investigation; funders and researchers should be encouraged not to turn their noses up at it.

Beyond people's having too much, or too little, disgust sensitivity, disgusting experiences can also cause psychological sequelae. Though post-traumatic stress disorder (PTSD) is generally thought of as being triggered by extreme fear, a companion condition is triggered by extreme disgust. Encountering decomposed corpses in war or at work, or other shockingly disgusting scenes, can lead to intrusive thoughts, flashbacks, recurrent nausea, and feelings of dirtiness that cannot be removed by washing—which can leave patients unable to lead a normal life.[58] Similarly, rape victims with PTSD can suffer from feelings of dirtiness, described as "mental pollution."[59] Victims of childhood sexual abuse and survivors of torture may also suffer persistent lasting mental torture by disgust.

Can anything be done to improve conditions for those suffering from disgust-related pathologies? I see good cause for hope. First, accurate diagnosis is required. Seeing such problems as part of an adaptive parasite-avoidance system can help hone the instruments of diagnosis and offer better means to test the effectiveness of therapies. Second, many of these conditions are part of a continuum in the population, with a somewhat arbitrary line dividing the "normal" from the "abnormal." Many people may be suffering at levels that are subclinical, but because the conditions are associated with shame, they may be reluctant to seek help. Health workers need to be briefed to detect hints of these conditions and to look for comorbidities, for example, for sexual dysfunction in those presenting with OCD. Internet-based support for conditions like these may provide part of the answer. Discussing such conditions online may be easier for sufferers than in a face-to-face interaction in a clinic.[60]

Third, there are many approaches to treatment, both through behavioral and drug therapies. A systematic look through the lens of disgust at what has worked in each of these conditions might reveal effective therapies. For example, we know that cognitive reappraisal is possible. Just as rotting milk can be relabeled as yogurt and so become palatable, exercises aimed at reappraising sexual organs, not as smelly and dirty, but as examples of exquisite design could be effective in reducing sexual phobias.[61] Work is needed to determine the effectiveness of behavioral therapies such as exposure with response prevention (ERP) and microbiological experiments demonstrating the lack of organisms on objects perceived to be contaminated.[62] Cognitive behavioral therapy (CBT) involving habituation to disgust objects, extinction of negative associations, and the formation of new and positive associations could be used across these phobias, possibly with the addition of cortisol, which has been shown to enhance the consolidation of newly learned memories.[63] If pathologies relate to disgust subsystems, it is vital that the right set of stimuli be used for habituation. Drug therapies might also focus on the possible implication of serotonin pathways in disgust.[64]

Finally, if I am right that different pathologies of the disgust system relate to its different components (sick people, hygiene, sex, food/animal, and fomite), then we have promising new avenues of research to help understand, and ultimately alleviate, these conditions. In appendix 1, our new disgust scale, which makes these distinctions, could prove a useful tool for this purpose.[65]

The Moral of Moral Disgust

We've looked for a moral to our disgust story in the microbe and manners domains, but what about in the moral domain? What can understanding the origins and purposes of moral disgust tell us that is useful?

First, our story reminds us that disgust, as one of the wellsprings of our sense of right and wrong, has an evolutionary history. It is an ancient system that arose in animals to help them avoid infection. This emotional system predates the evolution of "rationality" and still operates, to a large extent, independent of it. Though many prefer to believe that rationality is the source of our morality, evolutionary thinking has put paid to this idea.[66] Moral sentiments are just that—sentiments.

Emotional moral disgust underpins our very ability to live the ultrasocial human way of life. When we feel a flash of disgust at the individual who sneezes in our face in the elevator, or at the bully who pushes for an unfair share of the research fund, we are receiving messages from our ancestors telling us to act. They encourage us not to let wrongdoing, whether on a minor or major scale, pass. When we act on this disgust, we are behaving as moral beings, making social defectors pay the price of their defection. Each such disgust-driven act is a small contribution to the moral framework of social life.

The cognitive revolution of the 1960s undermined popular faith that it is right to act on our emotions. Rationality has become king. Professionals tell us that children should be given reasons to use a potty (nasty germs make you sick), when all the caregiver needs to do is to pull a disgusted face. Children or criminals who do wrong are counseled—ineffectually—about the consequences of their actions, when a simple dose of emotional learning via disgust and shame would be a surer way to set them on the right path.

Comparing the way I lived in traditional societies in Africa during my early professional career to the life I live now, among the UK's "chattering classes," I see a big difference in the place given to the moral emotions. There, I saw that those who did wrong were likely to meet with disgust and immediate social ostracism. Perhaps as a result, people seemed to pay more attention to social obligations, for example, always turning up for a wedding or funeral, than they do here. Here, I see folks dither about whether and how to respond to misdemeanors, apparently not trusting their emotions.

While trusting more to the emotions may have some upsides in terms of increased social cohesion in traditional societies, there are also downsides. Discrimination is a major problem for traditional values.[67] We know that disgust responses can be elicited by the sick, the homosexual, the sex worker, those with dirty jobs, and the social marginal or outsider.[68] Because moral disgust grew out of microbe disgust, does discrimination on the basis of potential infection leak into our moral systems?

In her book *Hiding for Humanity: Disgust, Shame and the Law?*, legal philosopher Martha Nussbaum argues that it does. She is suspicious of disgust as a moral emotion, saying that it engenders prejudice and stigmatization of the weak, the lower caste, the outsider, and the homosexual and that it can be used to contaminate the whole person.

Indeed, the unscrupulous and the power-hungry throughout history have long exploited this fact, whipping up minority, race, and out-group hatred so as to gain standing among the in-group as they close ranks.

Because we are hypervigilant to cues that might signal infectious substances or individuals, we pay attention when we are offered social information about potential disease threats. A common tactic for the playground bully is to label another child as infected or as having "cooties"; the victim then suffers shunning by their peer group. As the bullies grow up, they continue in the same vein, sometimes targeting not just individuals but whole groups. Ethnic strife abounds with the language of disgust; Tutsis are cockroaches, Jews lice, the "Japs" vermin, and the neighbors need ethnic "cleansing."[69] The bullies salve their consciences by thinking of their adversaries as "the great unwashed": dirty, degraded, and less than human. Robbed of claims to human dignity, the oppressed see their living conditions worsen: in slums, ghettos, camps, and marginal lands, hygiene becomes harder, fueling a vicious cycle of neglect and oppression.[70]

Throughout history, politicians have gained advantage by labeling members of out-groups as polluting, dirty, unhygienic, disease carriers. The rhetoric of disgust is called on to justify caste and class divisions, cruelty, exploitation, pogroms, ethnic cleansing, genocide, and war.[71] Such problems persist globally because the old tricks still work. The anti-gay politician or anti-abortion religious leader whips up disgust at an out-group; the implied disease threat then causes the in-group to close ranks. The effect is particularly visible at election times, when intercommunal violence peaks, as does the discussion of immigration.[72] By labeling the outsider as dirty and diseased, racists and nationalists find that they can also, to some extent, recruit morality to their side.[73] Those who campaign against abortion, homosexuality, and GM foods exploit the imagery and language of disgust and its ability to contaminate; they employ pictures of aborted fetuses, talk of "dirty" sexual practices, and raise the specter of "Frankenfoods."

This is the dangerous downside of disgust. But to entirely dismiss moral disgust as a basis for social justice, as Nussbaum suggests we do, is to throw the baby out with the bathwater. Leon Kass argues that there is wisdom in repugnance.[74] I would go further and say that without moral disgust of cheating, violence, and exploitation, we would be unable to function as a social species. None of the astonishing products of our modern supercooperative societies would be available to us, we would be plunged back into the Stone Age, and life would again become nasty, brutish, and short.

So the moral of moral disgust is that we need to listen to some kinds of disgust and not to others, and that we need science to help us to distinguish between them. The tricks of the politicians need to be exposed as such, and we all need access to the science that shows how our moral responses can be influenced by microbe disgust.

I have no doubt that such moral progress is possible.[75] Looking back at the list of what the kids in my local school found morally disgusting, the most frequently nominated category was discrimination against others, whether on the basis of race, gender, looks, age, religion, or sexual orientation. Over our history, the circle of what is regarded as the in-group has gradually grown from the clan, the tribe, and the nation to humanity as a whole, and even beyond.[76] This progress has come from expanded economic links, trade, and communication, from storytelling, to

TV, literature, and art—all products of the group imagination. Amazingly, for most of our history, humans probably thought of neighboring tribes as different species.[77] Science and art has now proved this to be wrong. We now know that we are all one. Strangers are no longer seen as less than human, to be disregarded, exploited, or treated as contaminated parasites. They are, instead, people like us.

EPILOGUE   DISGUST: THE UNFINISHED STORY

What does the future hold for disgust? The blooming field of disgustology is going to be an exciting place in the near future. We are not just going to learn a lot more about behavior relating to microbes, manners, and morality, but we will also blaze a trail for a new approach to brains and behavior. The implications are tremendous.

Psychology still has a long way to go before it can be regarded as a proper scientific discipline. The tools of psychological investigation are in their infancy; we've had the delicate probes and the huge rumbling scanners that can peer into living brains for only a couple of decades. Human brains are hard to study, as they are perhaps the most complex objects in existence. But these are only partial reasons for our slow progress.

In physics, the search for the Higgs boson began only after theoreticians had predicted that this fundamental particle should, of necessity, exist. The Large Hadron Collider was built to test a hypothesis that came out of theory. And theory is the right place to start in all of science. Patterns are noticed, a theory that explains the pattern is propounded, hypotheses are generated, and these are tested on new evidence. If the results do not fit the prediction, the hypothesis is wrong and so, probably, is the theory (or maybe the observations are mistaken—replication is important—either way).

Psychologists, however, distrust theory. Perhaps there have been too many wrong theories, like the work of Freud or Jung, or too many partial theories, like Pavlovian conditioning, cognitive dissonance, or the theory of reasoned action. Perhaps it is hard to theorize about brains because we are too close to them, and thinking about thinking about thinking soon becomes mind-bogglingly recursive. Or perhaps we've been too subjective, seeing human brains as special and therefore beyond theorizing.

The good news is that psychology is in the process of acquiring objective theory, and so is becoming a proper science. The theory that scientific psychology needs has been around for 150 years, but it has been misunderstood or has been rejected as simplistic or politically incorrect.[2] Evolution by natural selection provides our grand unifying theory for explaining brains and behavior.

In this book I've used evolutionary theory to propose a function for a class of behavior. Since parasites are a problem for all animals, infectious-disease-avoidance behavior must feature in the repertoire of all animals.[3] I found that the PAT of disgust neatly predicted the classes of behaviors that animals, including humans, engage in to avoid becoming prey to parasites. PAT gives us a simple, coherent, and parsimonious story about disgust that avoids some of the mental gymnastics employed by others who have tried to explain it.

The Disgust Agenda

But is PAT right? There is still a huge amount of research that we need to do to test the theory to destruction. What hypotheses can we test to check it out?

First, the story of animal infectious-disease avoidance needs completing. I predict that every species of fish, bird, or mammal will be found to avoid sick conspecifics, animals with infected lesions or signs of deformity, parasite eggs and cysts, and waste-contaminated foodstuffs (unless the benefits of contact outweigh the possible costs of infection). The epidemiology of infectious-disease transmission for different species can offer detailed predictions about their behavioral aversions.

If we look, I expect that we will also find that social animals will all display behaviors that help prevent parasite transmission. We should find that social animals have "manners" (meaning hygiene behaviors that reduce the likelihood of transmission of disease to others). Social animals, from insects to fish to birds to mammals, should display grooming and waste hygiene so as to make themselves more welcome as group members. They should sequester themselves when sick, and punish and/or avoid those who pose disease risk to others—the deformed, the sick, the incontinent, or the poorly groomed. Ethologists still have a lot of work to do to map disease-avoidance behavior in species other than our own.

Humans have a recently evolved prefrontal cortex that lets us metarepresent the world, model it in our heads, and learn from this virtual experience. Some argue that our special abilities have allowed us to dispense with our ancient animal "instincts." From the perspective of evolutionary theory, I predict that this "human exceptionalism" will be proved wrong. It will be established that humans come preadapted to (learn to) avoid poo, lesions, unhealthy mates, ectoparasites, and people who are unhygienic, and that many of the mechanisms by which we do these things are shared with the animals with which we have recent common ancestors. Eventually, when we've been able to pick apart the genes involved in disgust, it's my bet that DNA analysis will show that parts of our disgust system are deeply ancient, sharing much with reptile, or even invertebrate, aversion systems, because the tasks it has to do are similarly ancient.[4]

When it comes to human disgust mechanisms, there is so much still to learn. I expect that we will find that the neural basis for each of the subdomains of disgust is slightly different. Responses to stimuli concerning poor hygiene, lesions, deformity, sex, and contagion will share some neural mechanisms and have some that are different.

Another area that it will be possible to disentangle in the near future is the development of human disgust beyond disease avoidance. Disgust responses help avoid suboptimal mating—couplings with those of the "wrong" age or sex, those of poor mate quality, or with a high degree of relatedness. Incest avoidance disgust has received the most attention,[5] but suboptimal mating provides a simple explanation for why homosexual behavior is often reported as disgusting,[6] as is mating with someone of poor genetic quality (i.e., ugly) or displaying signs of sickness, or a young person having sex with an old person.[7] These "disgusts" should share features and brain mechanisms.

Other outgrowths of the disgust system will also be accounted for. In the next few years, we'll use fMRI to map disgust for outsiders and social defectors, and learn how much wiring belongs to the ancestral functions of disgust and how much is shared with its newer outgrowths. We'll figure out if moral disgust really is an outgrowth of microbe disgust, or something else entirely.

In short, we will soon be able to build a complete disgust map, telling the story of disgust in terms of function, mechanism, and behavior across species including our own. To do this correctly, we will have to let theory guide us, using the epidemiology of infectious disease as our guide.

Another area where we will see huge progress is in psychiatry, where we will understand and be able to treat the pathologies of the subdomains of the disgust system. I expect that overactive disgust for sex will be shown to cause problems with intimacy, overactive disgust for other people to cause social phobias, and overactive animal disgust to produce specific animal phobias. Disgust will also come to be more widely recognized as causing a distinct form of post-traumatic stress, which will be treated as such. Evolutionary thinking needs to blaze a trail into psychiatry,[8] giving us more effective chemical or cognitive alleviation for those with debilitating mental conditions.

Disgust as a Model Emotion

Disgust also has a lot to teach us about the emotions in general. Disgust's time has come, and where it leads, the other emotions will follow. The theoretical approach we've used to unweave disgust's rainbow works just as well for the other emotions. We can nail down thorny questions such as what emotions are for, how many emotions there are, how far we share them with related species, what subdomains they have, and ultimately how they actually work.[9]

If we want to use this evolutionary approach on other emotions, however, we'll have to leave some of our preconceptions at home. For example, if we want to understand love, we'd do better to start with the adaptive purposes of pair-bonding behavior rather than with how we feel around a significant other. We'll have to jettison common assumptions, especially the confusion of feelings with emotions. In the same way that disgust has an adaptive function, so too does love. Clear your head of that seductive, yearning, mushy feeling for the (wo)man in your life, and think instead of the mammals with offspring that are born so immature that they require huge parental investment to rear to adulthood. Accomplishing this task works best with two parents cooperating to rear the kids. But each parent might do better genetically if it left the other to do the job and went off to get more matings and more offspring. How to solve this problem? The only way is sacrifice. Each parent has to try to make it worth the other's while to stay around to help. Hence each offers what he or she can, paid in the currency of grooming services, food and shelter provisioning, sexual availability, entertainment, ego massage, and the cognitive economy of a regular, reliable domestic routine. The motive that makes each want to invest in the other in this way can be labeled "love."

Following the disgust example, we can make further predictions. We might expect that the love motive might be stronger in women than in men on average, because women are more heavily invested in their current children. We might expect that love would falter with the prospect of better reproductive options, and that social devices would have been invented to bolster the fragile pair-bond. In terms of mechanisms, we might expect to find a chemical arms race between males and females, with each trying to ensnare the other while remaining bond-free themselves.[10] And we might expect that individuals would be motivated to demonstrate their suitability as a potential mate with expensive displays of self-sacrifice, whether through risking disease by kissing and close contact, through buying expensive gifts, or through displaying their cooking abilities.

These types of behavior involve sacrifice of time and resource, but are necessary because the stakes in reproduction are so high. But why should we have *feelings* about love? Being able to reason consciously about the benefits and costs of any long-term plans is highly adaptive. If you can include what your ancient animal motives are telling you in this high-level cogitation, then you can make better plans. Hence a strong love *feeling* about someone is a signal between different parts of your brain; your lower-level motivated brain is drawing the attention of the reasoning brain to the fact that something important is going on down there, which you should pay attention to as you daydream about your long-term plans.

I could have written a book about love following exactly the same model as I have here for disgust. I'd have proposed its central adaptive function, traced the components of the behavioral tasks and their evolution in animals in different ecological niches. I'd have proposed how to disassociate distinct subcomponents of love according to the different types of problems that it typically meets, or the kinds of sacrificial behavior it motivates.[11] I could have described love as an adaptive system, varying by individuals due to innate factors and/or experience over the lifetime. I could have shown how love aids learning, with the reward system using signs of the happy other to tune the love behaviors of individuals. I could have looked at how love affects culture and how culture affects love. I could have investigated the pathologies of the love system, perhaps in religious fervor and stalking behavior. This alternative book about love would also have drawn conclusions about why love matters—lessons that could be important for human flourishing.

We can, and we need to, unpick love, and indeed all of the emotions, from the theoretical perspective of their ultimate evolutionary function, as we have done for disgust, if we are going to study emotion scientifically.

In the end, instead of writing about love, I chose to write a book about disgust. Intellectual curiosity was one reason; the other was my concern that this dark side of our emotional nature is neglected at our peril. As the instinct that protects us from microbes, it helps keep us healthy. As a factor behind manners, it helps us to be social, but also causes exclusion, stigma, and misery when it misfires. And as a component in our moral choices, disgust can hinder or help us build flourishing societies.

Writing this book has convinced me that figuring out how human behavior works, whether through the lens of disgust or of other motives such as love, could hardly be a more important task for us as a self-aware species. And I'm delighted to be living in a time when I can expect to see the next generation of scientists finally get to grips with what drives human behavior. One thing is certain: the story is only just beginning!

ACKNOWLEDGMENTS

This book benefited immeasurably from the wisdom of my collaborator and partner Robert Aunger. I would also like to thank Justin Aunger, Mícheál de Barra, Adam Biran, Sandy Cairncross, Sylvia Cremer, Dan Fessler, Joann Hoy, Jamie Large, Myriam Sidibé, Panos Tsagamoulis, Josh Tybur, and an anonymous reviewer for their advice. Christie Henry and Latha Menon ably and enthusiastically edited and steered the book to publication. My kids, Naïma and Abidine Sakandé, were an inspiration.

Chapter 6 is based on work previously published as V. Curtis, "Why Disgust Matters," *Philosophical Transactions of the Royal Society*, London B Biol Sci 2011; 366 (1583):3478–90.

APPENDIX:
THE LONDON DISGUST SCALE

		0	1	2	3	4	5	6
Directions: For each item, check a box to indicate how disgusted that situation makes you feel. 0 = no disgust, 3 = moderate disgust, 6 = extreme disgust									
	Food/Animal Disgust							
1	You crack open a boiled egg only to find a partially developed chick fetus inside.							
2	You are served a dish made of cow's tongue and cheek.							
3	Finding a dead mouse in the corner of your kitchen.							
4	You pour lumpy stale milk on your cereal.							
5	Seeing a cockroach run across your path.							
	Sex Disgust							
1	A street prostitute offers you sex for money.							
2	Shortly after meeting someone, you take them back to your house and have sex.							
3	Hearing about a woman who had sex with seven people in one day.							
4	You discover that your romantic partner once paid for sexual intercourse.							
5	A friend admits to attempting sexual intercourse with a piece of fruit.							
	Lesion Disgust							
1	You see a nurse dressing an infected wound; under the yellow bandages there is a weeping sore.							
2	Your friend shows you a big oozing lesion on his foot.							
3	On a medical TV program, you see some blisters on a male's genitals.							
4	Someone you work with develops a bad eye infection; the eye is almost fully sealed and weeps constantly.							
5	At a medical history museum, you see a wax model showing the effects of syphilis on the male and female body.							
	Atypical Appearance Disgust							
1	Sharing an elevator with a man with a disfigured face.							
2	Shaking hands with a homeless man.							
3	A woman with unkempt hair and disheveled clothes sits beside you on the bus.							
4	In a crowd you notice a man with one empty eye socket.							
5	Seeing an obese woman sunbathe.							

Hygiene Disgust								
1	Seeing some snotty tissues left on the table.							
2	You see someone sneeze phlegm onto their hands.							
3	Watching a woman pick her nose.							
4	On the subway, you are forced to stand close to someone with body odor and greasy hair.							
5	Feeling something sticky on a door handle.							
Fomite Disgust								
1	Without realizing, you use the dog's brush to brush your own hair.							
2	A stray dog licks you on your face.							
3	You accidentally use someone else's roll-on deodorant.							
4	A piece of toast drops butter-side down on the kitchen floor. You're hungry and it looks clean, so you pick it up and eat it anyway.							
5	At a restaurant, you notice you have accidentally been eating with a fork used by the person next to you.							

Scoring: Sum your ratings within each category to get your score for each kind of disgust. The possible category range is 0 to 30. Total disgust ranges from 0 to 180.

Results

Food/Animal Disgust   **3.5**

1. You crack open a boiled egg only to find a partially developed chick fetus inside.   4.2
2. You are served a dish made of cow's tongue and cheek.   3.8
3. Finding a dead mouse in the corner of your kitchen.   3.5
4. You pour lumpy stale milk on your cereal.   3.7
5. You see a cockroach run across your path.   2.5

Sex Disgust   **2.9**

1. A street prostitute offers you sex for money.   2.2
2. Shortly after meeting someone, you take them back to your house and have sex.   2.7
3. Hearing about a woman who had sex with seven people in one day.   3.1
4. You discover that your romantic partner once paid for sexual intercourse.   3.6
5. A friend admits to attempting sexual intercourse with a piece of fruit.   2.7

Lesion Disgust   **2.7**

1. You see a nurse dressing an infected wound; under the yellow bandages there
is a weeping sore.   2.6
2. Your friend shows you a big oozing lesion on his foot.   2.7
3. On a medical TV program, you see some blisters on a male's genitals.   3.0
4. Someone you work with develops a bad eye infection; the eye is almost fully sealed
and weeps constantly.   2.5
5. At a medical history museum, you see a wax model showing the effects of syphilis on the
male and female body.   2.7

Atypical Appearance Disgust   **1.8**

1. Sharing an elevator with a man with a disfigured face.   1.3
2. Shaking hands with a homeless man.   1.4
3. A woman with unkempt hair and disheveled clothes sits beside you on the bus.   2.2
4. In a crowd you notice a man with one empty eye socket.   2.0
5. Seeing an obese woman sunbathe.   1.9

Hygiene Disgust	**3.5**
1. Seeing some snotty tissues left on the table.	3.1
2. You see someone sneeze phlegm onto their hands.	3.9
3. Watching a woman pick her nose.	3.2
4. On the subway, you are forced to stand close to someone with body odor and greasy hair.	3.7
5. Feeling something sticky on a door handle.	3.5
Fomite Disgust	**2.6**
1. Without realizing, you use the dog's brush to brush your own hair.	2.6
2. A stray dog licks you on your face.	3.1
3. You accidentally use someone else's roll-on deodorant.	2.1
4. A piece of toast drops butter-side down on the kitchen floor. You're hungry and it looks clean, so you pick it up and eat it anyway.	2.3
5. At a restaurant, you notice you have accidentally been eating with a fork used by the person next to you.	3.0
Total	**2.8**

N = 550.

NOTES

Preface

1. Four books on disgust appeared in 2011/12: D. Kelly, *Yuck! The Nature and Moral Significance of Disgust* (Cambridge, MA: MIT Press, 2011); C. McGinn, *The Meaning of Disgust* (Oxford: Oxford University Press, 2011); R. Herz, *That's Disgusting: Unraveling the Mysteries of Repulsion* (New York: W. W. Norton and Company, 2012); C. Korsmeyer, *Savoring Disgust: The Foul and the Fair in Aesthetics* (Oxford: Oxford University Press, 2011). I disagree with most of the explanations for disgust that are found in these volumes. Other recent surveys of disgust include W. I. Miller, *The Anatomy of Disgust* (Cambridge, MA: Harvard University Press, 1997); and R. Rawdon Wilson, *The Hydra's Tale: Imagining Disgust* (Edmonton: University of Alberta Press, 2002). Two early twentieth-century studies of disgust are worth reading: A. Kolnai, *On Disgust*, trans. E. Kolnai (1927; Peru, IL: Carus, 2004); and W. Menninghaus, *Disgust: The Theory and History of a Strong Sensation* (New York; State University of New York Press, 2003).

2. The *New York Times* piece ("Survival's Ick Factor," by J. Gorman) can be found at http://www.nytimes.com/2012/01/24/science/disgusts-evolutionary -role-is-irresistible-to-researchers.html?pagewanted=all.

3. G. C. M. Jansen, A. O. Koloski-Ostrow, and E. M. Moormann, eds., *Roman Toilets: Their Archaeology and Cultural History* (Leuven, Belgium: Peeters Publishers, 2011); M. Bradley and K. Stow, eds., *Rome, Pollution and Propriety: Dirt, Disease and Hygiene in the Eternal City from Antiquity to Modernity* (Cambridge: Cambridge University Press, 2012).

4. J. M. Gottman, *What Predicts Divorce? The Relationship between Marital Processes and Marital Outcomes* (Hillsdale, NJ: Lawrence Erlbaum, 1994).

5. Paul Rozin and Jonathan Haidt did most of the pioneering scientific work on the topic of disgust. Their approach is summarized in P. Rozin, J. Haidt, and C. McCauley, "Disgust," in *Handbook of Emotion*, ed. M. Lewis, J. M. Haviland-Jones, and L. F. Barrett (New York: Guilford Press, 2008).

6. M. Douglas, *Purity and Danger: An Analysis of Concepts of Pollution and Taboo* (London: Routledge and Kegan Paul, 1966).

7. Robert Rawdon Wilson calls his book *The Hydra's Tale* in honor of disgust's many heads.

8. From the title of Richard Dawkins's book *Unweaving the Rainbow* (London: Allen Lane, 1998).

9. For brevity throughout this book, I use the words *parasite* and *pathogen* interchangeably. Biologically speaking, pathogens are a subcategory of parasites;

however, parasitologists and epidemiologists have more specific meanings for the terms, parasites tending to be multicellular organisms and pathogens being microbial and viral.

Chapter One

1. D. Heymann, ed., *Control of Communicable Diseases Manual*, 19th ed. (Washington, DC: American Public Health Association, 2008).

2. L. Liu et al., "Global, Regional, and National Causes of Child Mortality: An Updated Systematic Analysis for 2010 with Time Trends since 2000," *Lancet* 379 (2012).

3. A few authors prior to this had suggested that disease and disgust might be linked; for example, Randy Nesse and George C. Williams suggested that instinctive disgust keeps us from feces, vomit, and sources of contagion, and Steven Pinker called disgust "instinctive microbiology." R. M. Nesse and G. C. Williams, *Evolution and Healing* (London: Weidenfeld and Nicolson, 1995); S. Pinker, *How the Mind Works* (London: Penguin, 1998). Earlier than this, anthropologists such as Émile Durkheim and Marvin Harris attempted to explain the purification rituals and food taboos of different cultures as a sort of primitive hygiene. Mary Douglas labeled such thinking as "naïve medical materialism." Douglas, *Purity and Danger*. We were, however, first to provide a detailed account of the relationship between disease and disgust elicitors; see V. Curtis and A. Biran, "Dirt, Disgust, and Disease: Is Hygiene in Our Genes?" *Perspectives in Biology and Medicine* 44, no. 1 (2001); V. Curtis, M. de Barra, and R. Aunger, "Disgust as an Adaptive System for Disease Avoidance Behaviour," *Philosophical Transactions of the Royal Society B* 366 (2011).

4. These data, collected by master's student Panos Psagamoulis from a free listing of disgust items by passengers at Athens international airport, were published in Curtis and Biran, "Dirt, Disgust, and Disease."

5. It might be argued that the diseases I've cited are those that are found in medical textbooks today and are not the diseases that were around at the time that we evolved these aversions. Our modern human parasites are special in the sense that they have evolved in response to our recent shift to group living and animal husbandry, and they have evolved on a different trajectory in the time since we split from the great apes. But actually this is nothing special; evolutionary splits and changes in ecological circumstances are the story of all animals. All animals still have to deal with essentially the same pattern of viral, bacterial, fungal, protozoal, helminthic, and arthropodal parasites. Primates suffer from salmonellosis, giardiasis, candidiasis, malaria, schistosomiasis, hepatitis, TB, leprosy, Ebola, polio, ascariasis, pinworm, and amebiasis caused by the same, or cousin, agents to those causing human disease. Relations are close enough across primates for parasite cross-infection in both directions to have become a major issue for primatologists.

6. The use of words such as *instinctual* and *innate* have a long and contested

history. See M. Mameli and P. Bateson, "Innateness and the Sciences," *Biology and Philosophy* 21, no. 2 (2006), for a discussion. Previously I have suggested that disgust is an instinct that has to be learned—paraphrasing Mineka and Cook on the fear of snakes: S. Mineka and M. Cook, "Mechanisms Involved in the Observational Conditioning of Fear," *Journal of Experimental Psychology, General* 122 (1993). A more precise usage here might be "organized in advance of experience," a term coined by G. F. Marcus, *The Birth of the Mind: How a Tiny Number of Genes Creates the Complexities of Human Thought* (New York: Basic Books, 2004), which I took from J. Haidt, *The Righteous Mind: Why Good People Are Divided by Politics and Religion* (London: Penguin Books, 2010).

7. M. Fumagalli et al., "Signatures of Environmental Genetic Adaptation Pinpoint Pathogens as the Main Selective Pressure through Human Evolution," *PLOS Genetics* 7, no. 11 (2011).

8. Throughout this book I've referred to disgust and other motives as if they have agency, as if they "pull our strings" or tell us what to do. Some might argue that this is poetic license. Technically what is happening is that some aspect of the neurology of the brain causes an animal to bias its choices toward one course of action rather than another. These produce statistically discernible regularities in response to similar classes of stimuli in the environment, all else being equal. There is, of course, no actual voice or direct impulsion.

9. Curtis and Biran, "Dirt, Disgust, and Disease."

10. J. Haidt, C. McCauley, and P. Rozin, "Individual Differences in Sensitivity to Disgust: A Scale Sampling Seven Domains of Disgust Elicitors," *Personality and Individual Differences* 16, no. 5 (1994); P. Rozin et al., "Individual Differences in Disgust Sensitivity: Comparisons and Evaluations of Paper-and-Pencil versus Behavioral Measures," *Journal of Research in Personality* 33, no. 3 (1999).

11. A full account of Mícheál's research that resulted in our new disgust scale can be found in M. de Barra, "Attraction and Aversion: Pathogen Avoidance Strategies in the UK and Bangladesh" (PhD diss., London School of Hygiene and Tropical Medicine, 2011). Factor analysis is one of the best tools we have in psychology for pulling apart the mental mechanisms in the brain; however, the data it produces can leave room for interpretation. It looked as though, with a larger sample, the food and animal disgust might have broken into two factors, and the same might have happened for disease-risk sex, incest, and rule-breaking sex. This scale, which we've called the London Disgust Scale, can be found in the appendix.

12. Again see Mameli and Bateson, "Innateness and the Sciences."

13. E. Cashdan, "Adaptiveness of Food Learning and Food Aversions in Children," *Social Science Information* 37, no. 4 (1998).

14. R. M. Nesse, "Natural Selection and the Regulation of Defenses: A Signal Detection Analysis of the Smoke Detector Principle," *Evolution and Human Behavior* 26 (2005).

15. The field of disgust is still rife with such controversies.

16. See the excellent book on the history of controversies over sociobiology and evolutionary psychology by Kevin Laland and Gillian Brown, *Sense and Non-*

sense: Evolutionary Perspectives on Human Behaviour (Oxford: Oxford University Press, 2011). See also U. C. O. Segerstråle, *Defenders of the Truth: The Sociobiology Debate* (Oxford: Oxford University Press, 2001).

17. C. Darwin, *The Expression of the Emotions in Man and Animals* (1872; reprint, Chicago: University of Chicago Press, 1965).

18. Ibid.

19. This idea is was entrenched by the man who did the most for the science of disgust, Paul Rozin, who came to the topic via studies of taste aversions in rats. It is repeated in almost every work on disgust, for example in W. I. Miller, *Anatomy of Disgust*; S. B. Miller, *Disgust: The Gatekeeper Emotion* (Hillsdale, NJ: Analytic Press, 2004); Herz, *That's Disgusting*; Kelly, *Yuck!*

20. H. Spencer, *The Data of Ethics*, vol. 9 (London: Williams and Norgate, 1887), 206.

21. Mary Douglas was familiar with the Hindu purity-based caste system, which must have inspired her writing about the symbolic nature of dirt, though the book mostly concerns other cultures. Douglas invited me to speak at a seminar in honor of her eightieth birthday at University College, London, where she held to her view that disgusting matter was a product of cultural forces, while I said that it was a product of our biology, much older than society. I didn't win the debate with that audience of her devotees, but it was inspiring to see this sparkling figure wrestling with new ideas in her ninetieth decade.

22. S. Freud, J. Strachey, and A. Richards, *On Sexuality: Three Essays on the Theory of Sexuality and Other Works (1905)* (London: Penguin Books, 1977).

23. Kolnai, *On Disgust*.

24. A. Angyal, "Disgust and Related Aversions," *Journal of Abnormal and Social Psychology* 36 (1941).

25. Rozin's status and that of his work are pointed out in J. M. Tybur et al., "Disgust: Evolved Function and Structure," *Psychological Review* 120, no. 1 (2013).

26. Rozin, Haidt, and McCauley, "Disgust."

27. Rawdon Wilson, *Hydra's Tale*.

28. Curtis, de Barra, and Aunger, "Disgust as an Adaptive System."

Chapter Two

1. A. M. Kuris et al., "Ecosystem Energetic Implications of Parasite and Free-Living Biomass in Three Estuaries," *Nature* 454, no. 7203 (2008); C. Zimmer, *Parasite Rex* (New York: Touchstone, 2000).

2. We come equipped with most of the abilities of the animals in our ancestral phylogeny, and our brains betray this. See, for example, G. F. Streidter, *Principles of Brain Evolution* (Sunderland, MA: Sinauer Associates, 2005); J. Panksepp, *Affective Neuroscience* (Oxford: Oxford University Press, 1998). Of course, we are generalists with abilities to learn from our environments in

ways that way that outstrip our animal ancestors. But this does not mean that we have lost the ancient ancestral reflexive and motivational response systems: S. J. Shettleworth, "Modularity, Comparative Cognition and Human Uniqueness," *Philosophical Transactions of the Royal Society B* 367, no. 1603 (2012); R. Aunger and V. Curtis, "Kinds of Behaviour," *Biology and Philosophy* 23, no. 3 (2008).

3. Parasite-avoidance behavior in animals hasn't been a major focus of zoological investigation. See, for example, B. L. Hart, "Behavioural Defences in Animals against Pathogens and Parasites: Parallels with the Pillars of Medicine in Humans," *Philosophical Transactions of the Royal Society B* 366, no. 1583 (2011); S. Cremer, S. A. O. Armitage, and P. Schmid-Hempel, "Social Immunity," *Current Biology* 17, no. 16 (2007).

4. M. Kavaliers and D. D. Colwell, "Discrimination by Female Mice between the Odours of Parasitized and Non-Parasitized Males," *Proceedings of the Royal Society of London B* 261 (1995).

5. M. F. Spurier, M. S. Boyce, and B. F. J. Manly, "Effect of Parasites on Mate Choice of Captive Sage Grouse," in *Bird-Parasite Interactions: Ecology, Evolution, and Behaviour*, ed. J. E. Loye and M. Zuk (Oxford: Oxford University Press, 1991).

6. C. Loehle, "Social Barriers to Pathogen Transmission in Wild Animal Populations," *Ecology* 76 (1995).

7. W. J. Freeland, "Pathogens and the Evolution of Primate Sociality," *Biotropica* 8, no. 1 (1976).

8. R. Dawkins, *The Selfish Gene* (New York: Oxford University Press, 1976).

9. J. Keisecker et al., "Behavioral Reduction of Infection Risk," *Proceedings of the National Academy of Sciences USA* 96 (1999).

10. J. Krause and J.-G. J. Godin, "Influence of Parasitism on the Shoaling Behaviour of Banded Killifish, *Fundulus diaphanus*," *Canadian Journal of Zoology* 72 (1994).

11. D. C. Behringer, M. J. Butler, and J. D. Shields, "Avoidance of Disease by Social Lobsters," *Nature* 441 (2006).

12. Freeland, "Pathogens and the Evolution of Primate Sociality." Some doubt has been cast on these proposals by primatologists; the issue of primate social disease defense still requires definitive study.

13. D. W. Pfennig, M. L. G. Loeb, and J. P. Collins, "Pathogens as a Factor Limiting the Spread of Cannibalism in Tiger Salamanders," *Oecologia* 88 (1991).

14. M. Oaten, R. J. Stevenson, and T. I. Case, "Disease Avoidance as a Functional Basis for Stigmatization," *Philosophical Transactions of the Royal Society B* 366, no. 1583 (2011).

15. De Barra, "Attraction and Aversion."

16. H. Schulenburg and S. Muller, "Natural Variation in the Response of *Caenorhabditis elegans* towards *Bacillus thuringiensis*," *Parasitology* 128 (2004).

17. Cremer, Armitage, and Schmid-Hempel, "Social Immunity."

18. A. Karvonen, O. Seppala, and E. T. Valtonen, "Parasite Resistance and Avoidance Behaviour in Preventing Eye Fluke Infections in Fish," *Parasitology* 129, no. 2 (2004).

19. G. S. Wilkinson, "Social Grooming in the Common Vampire Bat, *Desmodus rotundus*," *Animal Behaviour* 34, no. 6 (1986).

20. M. S. Mooring, A. A. McKenzie, and B. L. Hart, "Grooming in Impala: Role of Oral Grooming in Removal of Ticks and Effects of Ticks in Increasing Grooming Rate," *Physiology and Behavior* 59 (1996).

21. P. R. Ehrlich, D. S. Dobkin, and D. Wheye, "The Adaptive Significance of Anting," *Auk* 103, no. 4 (1986).

22. J. T. Longino, "True Anting by the Capuchin, *Cebus capucinus*," *Primates* 25, no. 2 (1984); M. Zito, S. Evans, and P. J. Weldon, "Owl Monkeys (*Aotus* spp.) Self-Anoint with Plants and Millipedes," *Folia Primatologica* 74, no. 3 (2003).

23. R. Norval et al., "The Effect of the Bont Tick (*Amblyomma hebraeum*) on the Weight Gain of Africander Steers," *Veterinary Parasitology* 33, no. 3 (1989), cited in Mooring, McKenzie, and Hart, "Grooming in Impala."

24. C. B. Weddle, "Effects of Ectoparasites on Nestling Body Mass in the House Sparrow," *Condor* 102, no. 3 (2000).

25. M. D. Gumert, "Payment for Sex in a Macaque Mating Market " *Animal Behaviour* 74, no. 6 (2007).

26. D. H. Clayton, "Coevolution of Avian Grooming and Ectoparasite Avoidance," in *Bird-Parasite Interactions: Ecology, Evolution, and Behaviour*, ed. J. E. Loye and M. Zuk (Oxford: Oxford University Press, 1991).

27. S. J. O'Hara and P. C. Lee, "High Frequency of Post-Coital Penis Cleaning in Budongo Chimpanzees," *Folia Primatologica* 77, no. 5 (2006).

28. B. L. Hart, E. Korinek, and P. Brennan, "Postcopulatory Genital Grooming in Male Rats: Prevention of Sexually Transmitted Infections," *Physiology and Behavior* 41, no. 4 (1987).

29. M. Pagel and W. Bodmer, "A Naked Ape Would Have Fewer Parasites," *Biology Letters* 270, no. S1 (2003).

30. K. Norris, "A Trade-Off between Energy Intake and Exposure to Parasites in Oystercatchers Feeding on a Bivalve Mollusc," *Proceedings of the Royal Society of London B* 266, no. 1429 (1999).

31. G. S. Aeby, "Trade-Offs for the Butterflyfish, *Chaetodon multicinctus*, when Feeding on Coral Prey Infected with Trematode Metacercariae," *Behavioral Ecology and Sociobiology* 52 (2002).

32. S. Temple, "Do Predators Always Capture Substandard Individuals Disproportionately from Prey Populations?" *Ecology* 68, no. 3 (1987).

33. N. R. Franks et al., "Tomb Evaders: House-Hunting Hygiene in Ants," *Biology Letters* 1 (2005).

34. E. Diehl-Fleig and M. E. Lucchese, "Reações comportamentais de operárias de *Acromyrmex striatus* (Hymenoptera, Formicidae) na presença de fungos entomopatogênicos," *Revista Brasileira de Entomologia* 35 (1991). However, Sylva Cremer has just shown that sometimes ants will cluster around conspecifics infested with fungal pathogens and collect the spores, and so get the benefit of immunization. M. Konrad et al., "Social Transfer of Pathogenic Fungus Promotes Active Immunisation in Ant Colonies," *PLOS Biology* 10, no. 4 (2012).

35. E. Decaestecker, L. De Meester, and D. Ebert, "In Deep Trouble: Habitat Selection Constrained by Multiple Enemies in Zooplankton," *Proceedings of the National Academy of Sciences USA* 99, no. 8 (2002).

36. A. Oppliger, H. Richner, and P. Christie, "Effect of an Ectoparasite on Lay Date, Nest Site Choice, Desertion and Hatching Success in the Great Tit (*Parus major*)," *Behavioural Ecology* 5, no. 2 (1994).

37. S. A. Hosoi and S. I. Rothstein, "Nest Desertion and Cowbird Parasitism: Evidence for Evolved Responses and Evolutionary Lag," *Animal Behaviour* 59, no. 4 (2000).

38. M. R. Hutchings et al., "The Herbivores' Dilemma: Trade-Offs between Nutrition and Parasitism in Foraging Decisions," *Oecologia* 124 (2000).

39. Dung-based nematodes tend to increase egg laying during the spring calving. A. Gunn and R. J. Irvine, "Subclinical Parasitism and Ruminant Foraging Strategies: A Review," *Wildlife Society Bulletin* 31, no. 1 (2003). Other parasites, such as warble fly, have been shown to cause more of a problem the less migratory the host species: I. Folstad, O. Halvorsen, and A. C. Nilssen, "Parasite Avoidance: The Cause of Post-Calving Migrations in *Rangifer*?" *Canadian Journal of Zoology* 69, no. 9 (1991).

40. Martin Kavaliers, personal communication.

41. M. R. Weiss, "Good Housekeeping: Why Do Shelter-Dwelling Caterpillars Fling Their Frass?" *Ecology Letters* 6, no. 4 (2003).

42. J. E. Weaver and R. A. Sommers, "Life History and Habits of the Short-Tailed Cricket, *Anurogryllus muticus*, in Central Louisiana," *Annals of the Entomological Society of America* 62, no. 2 (1969).

43. V. G. Dethier, *The World of the Tent-Makers: A Natural History of the Eastern Tent Caterpillar* (Amherst: University of Massachusetts Press, 1980).

44. T. Eisner, E. van Tassell, and J. E. Carrel, "Defensive Use of a 'Fecal Shield' by a Beetle Larva," *Science* 158 (1967).

45. A. G. Hart and F. L. W. Ratnieks, "Task Partitioning, Division of Labour and Nest Compartmentalisation Collectively Isolate Hazardous Waste in the Leafcutting Ant *Atta cephalotes*," *Behavioral Ecology and Sociobiology* 49, no. 5 (2001); P. Schmid-Hempel, *Parasites in Social Insects* (Princeton, NJ: Princeton University Press, 1998).

46. M. J. West and R. D. Alexander, "Sub-Social Behavior in a Burrowing Cricket *Anurogryllus muticus* (De Geer)," *Ohio Journal of Science* 63 (1963); Y. Sato, Y. Saito, and T. Sakagami, "Rules for Nest Sanitation in a Social Spider Mite, *Schizotetranychus miscanthi* Saito (Acari: Tetranychidae)," *Ethology* 109, no. 9 (2003).

47. M. R. Pie, R. B. Rosengaus, and J. F. A. Traniello, "Nest Architecture, Activity Pattern, Worker Density and the Dynamics of Disease Transmission in Social Insects," *Journal of Theoretical Biology* 226, no. 1 (2004).

48. B. Hölldobbler and E. O. Wilson, *The Ants* (Cambridge, MA: Harvard University Press, 1990).

49. Ibid.

50. Cremer, Armitage, and Schmid-Hempel, "Social Immunity."

51. R. Krone et al., "Defecation Behaviour of the Lined Bristletooth Surgeon-fish *Ctenochaetus striatus* (Acanthuridae)," *Coral Reefs* 27, no. 3 (2008).

52. G. E. Brown, D. P. Chivers, and R. J. F. Smith, "Localized Defecation by Pike: A Response to Labelling by Cyprinid Alarm Pheromone?" *Behavioral Ecology and Sociobiology* 36, no. 2 (1995).

53. N. V. C. Polunin and I. Koike, "Temporal Focusing of Nitrogen Release by a Periodically Feeding Herbivorous Reef Fish," *Journal of Experimental Marine Biology and Ecology* 111, no. 3 (1987).

54. J. S. Brashares and P. Arcese, "Scent Marking in a Territorial African Antelope: II. The Economics of Marking with Faeces," *Animal Behaviour* 57, no. 1 (1999).

55. There are examples of all these behaviors in multiple species of birds. See, for example, R. H. Blair and B. W. Tucker, "Nest Sanitation," *British Birds* 34 (1941); W. A. Montevecchi, "Eggshell Removal by Laughing Gulls," *Bird-Banding* 47, no. 2 (1976); S. Hurtrez-Boussès et al., "Effects of Ectoparasites of Young on Parents' Behaviour in a Mediterranean Population of Blue Tits," *Journal of Avian Biology* 31, no. 2 (2000); H. Mayfield, "Nesting Success Calculated from Exposure," *Wilson Bulletin* (1961); A. Mennerat et al., "Aromatic Plants in Nests of the Blue Tit *Cyanistes caeruleus* Protect Chicks from Bacteria," *Oecologia* 161, no. 4 (2009); F. R. Gehlbach and R. S. Baldridge, "Live Blind Snakes (*Leptotyphlops dulcis*) in Eastern Screech Owl (*Otus asio*) Nests: A Novel Commensalism," *Oecologia* 71, no. 4 (1987).

56. V. B. Meyer-Rochow and J. Gal, "Pressures Produced When Penguins Pooh: Calculations on Avian Defaecation," *Polar Biology* 27 (2003).

57. P. T. Fretwell and P. N. Trathan, "Penguins from Space: Faecal Stains Reveal the Location of Emperor Penguin Colonies," *Global Ecology and Biogeography* 18, no. 5 (2009).

58. These swallows can be seen at http://www.youtube.com/watch?v=Fa5 CluKJxGM.

59. C. Petit et al., "Blue Tits Use Selected Plants and Olfaction to Maintain an Aromatic Environment for Nestlings," *Ecology Letters* 5 (2002).

60. P. H. Wimberger, "The Use of Green Plant Material in Bird Nests to Avoid Ectoparasites," *Auk* 101, no. 3 (1984); L. Clark and J. R. Mason, "Use of Nest Material as Insecticidal and Anti-Pathogenic Agents by the European Starling," *Oecologia* 67, no. 2 (1985).

61. Hurtrez-Boussès et al., "Effects of Ectoparasites of Young"; J. Banbura et al., "Sex Differences in Parental Care in a Corsican Blue Tit *Parus caeruleus* Population," *Ardea* 89, no. 3 (2001).

62. Gehlbach and Baldridge, "Live Blind Snakes."

63. The word *parasite* itself—which comes from *para-sitos*, meaning "beside-bread"—has its origins in the custom of ancient Greece. Priests shared their food at temples; those who relied on such handouts came to be seen as hangers-on and welfare cheats. Robert Parker, personal communication.

64. J. Garcia, D. Forthman-Quick, and B. White, "Conditioned Disgust

and Fear from Mollusk to Monkey," in *Primary Neural Substrates of Learning and Behavioral Change* (New York: Cambridge University Press, 1984), 57.

65. A common use of serotonin in avoidance behavior across animals is possible. M. Rubio-Godoy, R. Aunger, and V. Curtis, "Serotonin: A Link between Disgust and Immunity?" *Medical Hypotheses* 68, no. 1 (2007).

66. The standard model of disgust seems to imply this, e.g., Rozin, Haidt, and McCauley, "Disgust."

Chapter Three

1. D. W. Pfennig, S. G. Ho, and E. A. Hoffman, "Pathogen Transmission as a Selective Force against Cannibalism," *Animal Behaviour* 55, no. 5 (1998).

2. J. B. H. Savigny and A. Corréard, *Narrative of a Voyage to Senegal in 1816: Undertaken by Order of the French Government, Comprising an Account of the Shipwreck of the Medusa, the Sufferings of the Crew, and the Various Occurrences on Board the Raft, in the Desert of Zaara, at St. Louis, and at the Camp of Daccard . . .* (London: H. Colburn, 1818).

3. The big five dimensions of personality are openness, conscientiousness, extraversion, agreeableness, and neuroticism. J. Hennig, P. Pössel, and P. Netter, "Sensitivity to Disgust as an Indicator of Neuroticism: A Psychobiological Approach," *Personality and Individual Differences* 20, no. 5 (1996). But for a differing view, see J. M. Tybur et al., "Sex Differences and Sex Similarities in Disgust Sensitivity," *Personality and Individual Differences* 51, no. 3 (2011).

4. D. Nettle, *Personality: What Makes You the Way You Are* (Oxford: Oxford University Press, 2007).

5. D. Nettle, "The Evolution of Personality Variation in Humans and Other Animals," *American Psychologist* 61, no. 6 (2006).

6. M. Wolf and F. J. Weissing, "Animal Personalities: Consequences for Ecology and Evolution," *Trends in Ecology and Evolution* 27, no. 8 (2012).

7. A. Sih, A. Bell, and J. C. Johnson, "Behavioral Syndromes: An Ecological and Evolutionary Overview," *Trends in Ecology and Evolution* 19, no. 7 (2004).

8. V. Curtis, R. Aunger, and T. Rabie, "Evidence That Disgust Evolved to Protect from Risk of Disease," *Proceedings of the Royal Society B* 271, supplement 4 (2004).

9. Rozin, Haidt, and McCauley, "Disgust."

10. See Unilever's award-winning ad for Lynx deodorant at http://www.youtube.com/watch?v=ensckApupW0, for example.

11. D. M. T. Fessler, S. J. Eng, and C. D. Navarrete, "Elevated Disgust Sensitivity in the First Trimester of Pregnancy: Evidence Supporting the Compensatory Prophylaxis Hypothesis," *Evolution and Human Behavior* 26 (2005).

12. S. M. Flaxman and P. W. Sherman, "Morning Sickness: A Mechanism for Protecting Mother and Embryo," *Quarterly Review of Biology* 75, no. 2 (2000); G. V. Pepper and S. C. Roberts, "Rates of Nausea and Vomiting in Pregnancy and Dietary Characteristics across Populations," *Proceedings of the Royal Society B* 273, no. 1601 (2006).

13. D. S. Fleischman and D. M. T. Fessler, "Progesterone's Effects on the Psychology of Disease Avoidance: Support for the Compensatory Behavioral Prophylaxis Hypothesis," *Hormones and Behavior* 59, no. 2 (2011).

14. M. Schaller et al., "Mere Visual Perception of Other People's Disease Symptoms Facilitates a More Aggressive Immune Response," *Psychological Science* 21, no. 5 (2010).

15. Rubio-Godoy, Aunger, and Curtis, "Serotonin: A Link between Disgust and Immunity?"

16. O. Curno et al., "Mothers Produce Less Aggressive Sons with Altered Immunity When There Is a Threat of Disease during Pregnancy," *Proceedings of the Royal Society of London B* 276, no. 1659 (2009).

17. A. Hoefling et al., "When Hunger Finds No Fault with Moldy Corn: Food Deprivation Reduces Food-Related Disgust," *Emotion* 9, no. 1 (2009).

18. D. Ariely, *Predictably Irrational: The Hidden Forces That Shape Our Decisions* (London: HarperCollins, 2008).

19. Otherwise known as the learned taste aversion.

20. Garcia, Forthman-Quick, and White, "Conditioned Disgust and Fear." Pond snails also acquire differential conditioned taste aversion to either sucrose or carrot juice paired with lithium chloride (LiCl). R. Sugai et al., "Taste Discrimination in Conditioned Taste Aversion of the Pond Snail *Lymnaea stagnalis*," *Journal of Experimental Biology* 209, no. 5 (2006).

21. M. E. P. Seligman, "Phobias and Preparedness," *Behavior Therapy* 2, no. 3 (1971).

22. R. Soussignan et al., "Facial and Autonomic Responses to Biological and Artificial Olfactory Stimuli in Human Neonates: Re-examining Early Hedonic Discrimination of Odors," *Physiology and Behavior* 62, no. 4 (1997); Y. Yeshurun and N. Sobel, "An Odor Is Not Worth a Thousand Words: From Multidimensional Odors to Unidimensional Odor Objects," *Annual Review of Psychology* 61 (2010).

23. At the recent disgust conference in Bielefeld, I learned from Martin Kavaliers, who studies aversions in rodents, that mice tend to avoid mice who have been in contact with sick mice. This is close to the human ability to detect contamination using the memory of what has been in contact with what, but a chemical pathway can't be ruled out.

24. R. F. Baumeister, K. D. Vohs, and C. Nathan DeWall, "How Emotion Shapes Behavior: Feedback, Anticipation, and Reflection, Rather Than Direct Causation," *Personality and Social Psychology Review* 11, no. 2 (2007).

25. G. C. L. Davey et al., "A Cross-Cultural Study of Animal Fears," *Behaviour Research and Therapy* 36 (1998); B. O. Olatunji, J. P. Forsyth, and A. Cherian, "Evaluative Differential Conditioning of Disgust: A Sticky Form of Relational Learning That Is Resistant to Extinction," *Journal of Anxiety Disorders* 21 (2007).

26. S. Rachman, "Fear of Contamination," *Behaviour Research and Therapy* 42 (2004).

27. D. F. Tolin, P. Worhunsky, and N. Maltby, "Sympathetic Magic in Contam-

ination-Related OCD," *Journal of Behavior Therapy and Experimental Psychiatry* 35, no. 2 (2004).

28. I will discuss controversies about group selection in chapter 6.

29. This idea is highly controversial at present. Two recent additions to the debate are from Steven Pinker, who does not subscribe to group selection as a force in human evolution: "The False Allure of Group Selection," *Edge*, http://edge.org/conversation/the-false-allure-of-group-selection (2012); and E. O. Wilson, who champions it: *The Social Conquest of Earth* (New York: Liveright Publishing, 2012).

30. S. Lindenbaum, "Understanding Kuru: The Contribution of Anthropology and Medicine," *Philosophical Transactions of the Royal Society B* 363, no. 1510 (2008); D. C. Gajdusek, "Unconventional Viruses and the Origin and Disappearance of Kuru," in *Nobel Lectures in Physiology or Medicine, 1971–1980*, ed. J. Lindsten (Stockholm: Karolinska Institute, 1977).

31. G. Cochran and H. Harpending, *The 10,000 Year Explosion: How Civilization Accelerated Human Evolution* (New York: Basic Books, 2010).

32. Thanks to master gamer Justin Aunger for providing the gory details.

33. W. Arens, *The Man-Eating Myth: Anthropology and Anthropophagy* (Oxford: Oxford University Press, 1979).

34. J. M. Diamond, "Archaeology: Talk of Cannibalism," *Nature* 407, no. 6800 (2000).

35. R. A. Marlar et al., "Biochemical Evidence of Cannibalism at a Prehistoric Puebloan Site in Southwestern Colorado," *Nature* 407, no. 6800 (2000).

36. For a more complete account of disgust as an adaptive system, see Curtis, de Barra, and Aunger, "Disgust as an Adaptive System."

Chapter Four

1. S. L. Carter, *Civility: Manners, Morals, and the Etiquette of Democracy* (New York: Basic Books, 1998).

2. Norbert Elias's book *The Civilizing Process* (1939; trans E. Jephcott [Malden, MA: Blackwell, 2000]) made the biggest contribution, suggesting that changing social attitudes molded standards of behavior regarding bodily functions, clothing, and forms of speech from the courtly behavior of the late Middle Ages to the present day. The structuralist sociologist Pierre Bourdieu ("Social Space and Symbolic Power," *Sociological Theory* 7, no. 1 [1989]) has manners a part of the human *habitus*, the taken-for-granted, tacit codes of conduct into which we are socialized and which serve primarily to symbolize and signal our social status. Following this line of thought, Michèle Lamont's book *Money, Morals, and Manners* (Chicago: University of Chicago Press, 1992) explored how the upper classes in the United States and France distinguish themselves as a group apart: "High-class people despise vulgarity and bad manners and greatly appreciate charm and intelligence" (167). However, Lamont does not attempt to explain manners.

Nor does Claude Lévi-Strauss in the volume of his magnum opus *Mythologiques* entitled *The Origin of Table Manners* (Chicago: University of Chicago Press, 1990).

3. R. Wrangham, *Catching Fire: How Cooking Made Us Human* (London: Profile Books, 2010).

4. S. Mithen, *The Prehistory of the Mind: A Search for the Origins of Art, Religion and Science* (London: Thames and Hudson, 1996).

5. M. Ridley, *The Rational Optimist: How Prosperity Evolves* (London: Fourth Estate, 2010); M. Tomasello and M. Carpenter, "Shared Intentionality," *Developmental Science* 10, no. 1 (2007); K. N. Laland, N. Atton, and M. M. Webster, "From Fish to Fashion: Experimental and Theoretical Insights into the Evolution of Culture," *Philosophical Transactions of the Royal Society B* 366, no. 1567 (2011); S. Bowles and H. Gintis, *A Cooperative Species: Human Reciprocity and Its Evolution* (Princeton, NJ: Princeton University Press, 2011).

6. Freeland, "Pathogens and the Evolution of Primate Sociality."

7. There are many proposals for the functional basis of shame in the literature, none of which fit it to disgust and manners in quite this way. Closest is Dan Fessler, "Shame in Two Cultures: Implications for Evolutionary Approaches," *Journal of Cognition and Culture* 4, no. 2 (2004). Roger Giner-Sorolla's latest study also supports the idea of shame as a corollary to disgust; R. Giner-Sorolla and P. Espinosa, "Social Cuing of Guilt by Anger and of Shame by Disgust," *Psychological Science* 22, no. 1 (2011).

8. R. J. Stevenson et al., "Children's Response to Adult Disgust Elicitors: Development and Acquisition," *Developmental Psychology* 46, no. 1 (2010).

9. V. Curtis et al., "Potties, Pits and Pipes: Explaining Hygiene Behaviour in Burkina Faso," *Social Science and Medicine* 41, no. 3 (1995).

10. V. Curtis, "Dirt, Disgust and Disease: A Natural History of Hygiene," *Journal of Epidemiology and Community Health* 61, no. 8 (2007); V. S. Smith, *Clean: A History of Personal Hygiene and Purity* (Oxford: Oxford University Press, 2007).

11. D. E. Brown, *Human Universals* (New York: McGraw-Hill, 1991).

12. De Barra, "Attraction and Aversion."

13. N. Elias, *The Civilizing Process*, translated by E. Jephcott (1939; Malden, MA: Blackwell, 2000).

14. Though not all ideas that get passed down are necessarily beneficial. R. Aunger, "Are Food Avoidances Maladaptive in the Ituri Forest of Zaire?" *Journal of Anthropological Research* 50 (1994).

15. S. Mathew and R. Boyd, "Punishment Sustains Large-Scale Cooperation in Prestate Warfare," *Proceedings of the National Academy of Sciences USA* 108, no. 28 (2011).

16. J. Henrich, "The Evolution of Costly Displays, Cooperation and Religion: Credibility Enhancing Displays and Their Implications for Cultural Evolution," *Evolution and Human Behavior* 30, no. 4 (2009).

17. S. Nichols, "On the Genealogy of Norms: A Case for the Role of Emotion in Cultural Evolution," *Philosophy of Science* 69 (2002).

18. Elias, *Civilizing Process*; A. J. Packer and P. Espeland, *How Rude! The Teenagers' Guide to Good Manners, Proper Behavior, and Not Grossing People Out* (Minneapolis, MN: Free Spirit Publishing, 1997).

19. William of Wykeham (1324–1404), bishop of Winchester and chancellor of England, was the founder of Winchester College and New College, Oxford. "Maner mayks man" (c. 1350 Douce MS 52 no. 77), *Oxford Dictionary of Proverbs* website.

Chapter Five

1. A. Smith, *Wealth of Nations* (Wiley Online Library, 1999).

2. E. J. Chaisson, "Energy Rate Density as a Complexity Metric and Evolutionary Driver," *Complexity* 16, no. 3 (2011).

3. W. Hamilton, "The Genetical Evolution of Social Behaviour," *Journal of Theoretical Biology* 7 (1964).

4. R. L. Trivers, "The Evolution of Reciprocal Altruism," *Quarterly Review of Biology* 46 (1971).

5. R. Joyce, *The Evolution of Morality* (Cambridge, MA: MIT Press, 2006).

6. M. A. Nowak and K. Sigmund, "Evolution of Indirect Reciprocity," *Nature* 437, no. 7063 (2005).

7. W. H. Durham, *Coevolution: Genes, Culture, and Human Diversity* (Stanford, CA: Stanford University Press, 1991).

8. Ibid.

9. E. Sober and D. S. Wilson, *Unto Others: The Evolution and Psychology of Unselfish Behavior* (Cambridge, MA: Harvard University Press, 1999).

10. M. Nowack and R. Highfield, *Supercooperators: Evolution, Altruism and Human Behaviour; or, Why We Need Each Other to Succeed* (New York: Free Press, 2011); Bowles and Gintis, *A Cooperative Species*. But also see Steven Pinker's piece "False Allure of Group Selection."

11. *We Need to Talk about Kevin*, directed by Lynne Ramsay, 2011.

12. T. Wheatley and J. Haidt, "Hypnotic Disgust Makes Moral Judgments More Severe," *Psychological Science* 16, no. 10 (2005).

13. S. Schnall et al., "Disgust as Embodied Moral Judgment," *Personality and Social Psychology Bulletin* 34, no. 8 (2008).

14. S. Schnall, J. Benton, and S. Harvey, "With a Clean Conscience," *Psychological Science* 19, no. 12 (2008).

15. C. B. Zhong and K. Liljenquist, "Washing Away Your Sins: Threatened Morality and Physical Cleansing," *Science* 313, no. 5792 (2006).

16. J. V. Fayard et al., "Is Cleanliness Next to Godliness? Dispelling Old Wives' Tales: Failure to Replicate Zhong and Liljenquist (2006)," *Journal of Articles in Support of the Null Hypothesis* 6, no. 2 (2009). People conditioned to feel disgust at certain words also didn't show any more moral disgust when these words were incorporated into scenarios of moral wrongdoing: B. David and B. O. Olatunji, "The Effect of Disgust Conditioning and Disgust Sensitivity on Appraisals of Moral Transgressions," *Personality and Individual Differences* 50, no. 7 (2011). Other disgust researchers have told me that they couldn't replicate the Wheatley and Haidt and Schnall findings. More studies are still coming in: e.g., Z. Yan, D. Ding, and L. Yan, "To Wash Your Body, or Purify Your Soul: Physical Cleansing

Would Strengthen the Sense of High Moral Character," *Psychology* 2, no. 9 (2011); K. J. Eskine, N. A. Kacinik, and J. J. Prinz, "A Bad Taste in the Mouth," *Psychological Science* 22, no. 3 (2011); K. Liljenquist, C. B. Zhong, and A. D. Galinsky, "The Smell of Virtue: Clean Scents Promote Reciprocity and Charity," *Psychological Science* 21, no. 3 (2010).

17. H. A. Chapman et al., "In Bad Taste: Evidence for the Oral Origins of Moral Disgust," *Science* 323, no. 5918 (2009).

18. E. Royzman and R. Kurzban, "Minding the Metaphor: The Elusive Character of Moral Disgust," *Emotion Review* 3, no. 3 (2011).

19. Stevenson et al., "Children's Response to Adult Disgust Elicitors."

20. Giner-Sorolla and Espinosa, "Social Cuing of Guilt."

21. Martin Kavaliers, personal communication.

22. J. Moll et al., "The Moral Affiliations of Disgust: A Functional MRI Study," *Cognitive and Behavioral Neurology* 18, no. 1 (2005): 68–78.

23. A. G. Sanfey et al., "The Neural Basis of Economic Decision-Making in the Ultimatum Game," *Science* 300, no. 5626 (2003).

24. We haven't specifically examined the question of incest disgust. Debra Lieberman and her colleagues would argue that this is another special category of moral disgust, one that may originate from the adaptive need to avoid inbreeding: D. Lieberman, J. Tooby, and L. Cosmides, "The Architecture of Human Kin Detection," *Nature* 445, no. 7129 (2007).

25. Our view of the evolved basis of the control of behavior and the role of emotions is set out more fully in Aunger and Curtis, "Kinds of Behaviour." It is also the subject of a forthcoming book with Oxford University Press.

26. A. R. Damasio, *Descartes' Error: Emotion, Reason, and the Human Brain* (New York: Putnam, 1994).

Chapter Six

1. Of course, what is meant by "better" is a topic for philosophers. Some say we can't describe those higher values, but I'm in the camp that thinks that we can and should do so, as is Sam Harris: *The Moral Landscape: How Science Can Determine Human Values* (New York: Free Press, 2011).

2. UNICEF and World Health Organization, "Diarrhoea: Why Children Are Still Dying and What Can Be Done" (New York: UNICEF, 2009); Liu et al., "Causes of Child Mortality."

3. World Health Organization and UNICEF, *Progress on Sanitation and Drinking-Water: 2010 Update* (Geneva: World Health Organization, 2010).

4. Our original figure was a million children, which we have revised downward because of the global fall in infant mortality. K. Greenland et al., "Editorial: Can We Afford to Overlook Hand Hygiene Again?" *Tropical Medicine and International Health* 18, no. 3 (2013); V. Curtis and S. Cairncross, "Effect of Washing Hands with Soap on Diarrhoea Risk in the Community: A Systematic Review," *Lancet Infectious Diseases* 3 (2003).

5. V. Curtis et al., "Hygiene: New Hopes, New Horizons," *Lancet Infectious Diseases* 11, no. 4 (2011).

6. T. Rabie and V. Curtis, "Handwashing and Risk of Respiratory Infections: A Quantitative Systematic Review," *Tropical Medicine and International Health* 11, no. 3 (2006); I. C.-H. Fung and S. Cairncross, "Effectiveness of Handwashing in Preventing SARS: A Review," *Tropical Medicine and International Health* 11, no. 11 (2006); T. Jefferson et al., "Physical Interventions to Interrupt or Reduce the Spread of Respiratory Viruses: Systematic Review," *British Medical Journal* 336 (2008); P. M. Emerson et al., "A Review of the Evidence for the 'F' and 'E' Components of the SAFE Strategy for Trachoma Control," *Tropical Medicine and International Health* 5, no. 8 (2000); D. B. Huang and J. Zhou, "Effect of Intensive Handwashing in the Prevention of Diarrhoeal Illness among Patients with AIDS: A Randomized Controlled Study," *Journal of Medical Microbiology* 56, no. 5 (2007); J. H Humphrey, "Child Undernutrition, Tropical Enteropathy, Toilets, and Handwashing," *Lancet* 374 (2009).

7. V. A. Curtis et al., "Hygiene in the Home: Relating Bugs to Behaviour," *Social Science and Medicine* 57, no. 4 (2003).

8. V. Curtis, L. Danquah, and R. Aunger, "Planned, Motivated and Habitual Hygiene Behaviour: An Eleven Country Review," *Health Education Research* 24, no. 4 (2009).

9. Curtis, Danquah, and Aunger, "Planned, Motivated and Habitual Hygiene Behaviour."

10. B. E. Scott et al., "Marketing Hygiene Behaviours: The Impact of Different Communications Channels on Reported Handwashing Behaviour of Women in Ghana," *Health Education Research* 22, no. 4 (2007).

11. The ad can be seen at http://www.youtube.com/watch?v=w2qRcMTstzc.

12. G. Judah et al., "Experimental Pretesting of Hand-Washing Interventions in a Natural Setting," *American Journal of Public Health* 99, no. S2 (2009).

13. NHS website, http://www.dh.gov.uk/prod_consum_dh/groups/dh_digitalassets/@dh/@en/documents/digitalasset/dh_098680.pdf.

14. G. J. Rubin, H. W. W. Potts, and S. Michie, "The Impact of Communications about Swine Flu (Influenza H1N1v) on Public Responses to the Outbreak: Results from 36 National Telephone Surveys in the UK," *Health Technology Assessment* 14, no. 34 (2010).

15. D. S. Fleischman et al., "Sensor Recorded Changes in Rates of Hand Washing with Soap in Response to the Media Reports of the H1N1 Pandemic in Britain," *BMJ Open* 1 (2011).

16. A. Brian et al., "Superamma: A Cluster Randomized Trial of a Village-Level Intervention to Promote Handwashing with Soap in Rural Andhra Pradesh, India" (unpublished manuscript, 2013).

17. BBC website, http://www.bbc.co.uk/news/world-asia-india-20202339.

18. Golden Poo Awards, http://www.thegoldenpooawards.org/.

19. British Heart Foundation, "Give Up before You Clog Up," edited by UK Department of Health, advertisement on YouTube, http://www.youtube.com/watch?v=cDAN7Oi62e0, 2003.

20. D. Hammond et al., "Graphic Canadian Cigarette Warning Labels and Adverse Outcomes: Evidence from Canadian Smokers," *American Journal of Public Health* 94, no. 8 (2004).

21. G. Gilbert, director, *Jamie's School Dinners* (Channel 4 [UK], 2005).

22. New York City Department of Health and Mental Hygiene, http://www .youtube.com/watch?v=Omt-i2aypew.

23. Both fear and disgust operate on the smoke-alarm principle, better to be safe than sorry. In the case of fear, it is adaptive to overdetect snakes in the grass on average, and so waste energy avoiding a harmless stick, as the rare snakebite is extremely costly. In the same way it is better, on average, to avoid food that smells a bit off, even if it means missing a meal, than it is to eat everything and perhaps die from typhoid. See Nesse, "Natural Selection and the Regulation of Defenses"; M. G. Haselton and D. Nettle, "The Paranoid Optimist: An Integrative Evolutionary Model of Cognitive Biases," *Personality and Social Psychology Review* 10, no. 1 (2006).

24. C .W. Schmidt, "The Yuck Factor: Where Disgust Meets Discovery," *Environmental Health Perspective* 116, no. 12 (2008).

25. S. Earle, "Factors Affecting the Initiation of Breastfeeding: Implications for Breastfeeding Promotion," *Health Promotion International* 17, no. 3 (2002).

26. A. Corbin, *The Foul and the Fragrant: Odor and the French Social Imagination* (Cambridge, MA: Harvard University Press, 1988).

27. Herz, *That's Disgusting*.

28. W. Marsiglio, "Attitudes toward Homosexual Activity and Gays as Friends: A National Survey of Heterosexual 15- to 19-Year-Old Males," *Journal of Sex Research* 30, no. 1 (1993).

29. R. Kurzban and M. R. Leary, "Evolutionary Origins of Stigmatization: The Functions of Social Exclusion," *Psychological Bulletin* 127, no. 2 (2001).

30. J. H. Park, J. Faulkner, and M. Schaller, "Evolved Disease-Avoidance Processes and Contemporary Anti-Social Behavior: Prejudicial Attitudes and Avoidance of People with Physical Disabilities," *Journal of Nonverbal Behavior and Brain Sciences* 27, no. 2 (2003); J. H. Park, M. Schaller, and C. S. Crandall, "Pathogen-Avoidance Mechanisms and the Stigmatization of Obese People," *Evolution and Human Behavior* 28, no. 6 (2007).

31. Park, Faulkner, and Schaller, "Evolved Disease-Avoidance Processes"; Park, Schaller, and Crandall, "Pathogen-Avoidance Mechanisms"; L. A. Duncan and M. Schaller, "Prejudicial Attitudes toward Older Adults May Be Exaggerated When People Feel Vulnerable to Infectious Disease: Evidence and Implications," *Analyses of Social Issues and Public Policy* 9, no. 1 (2009).

32. Curtis, Aunger, and Rabie, "Evidence That Disgust Evolved."

33. P. Magin et al., "Psychological Sequelae of Acne Vulgaris: Results of a Qualitative Study," *Canadian Family Physician* 52, no. 8 (2006); A. P. Alio et al., "The Psychosocial Impact of Vesico-Vaginal Fistula in Niger," *Archives of Gynecology and Obstetrics* 284 (2011).

34. R. Dahle, "Lukt, fukt og skam" (Bodily smells, wetness and shame; paper presented at Norsk Selskap for Aldersforskning, Oslo, September 9, 1999).

35. L. W. Isaksen, "Toward a Sociology of (Gendered) Disgust Images of

Bodily Decay and the Social Organization of Care Work," *Journal of Family Issues* 23, no. 7 (2002).

36. J. Wrubel and S. Folkman, "What Informal Caregivers Actually Do: The Caregiving Skills of Partners of Men with AIDS," *AIDS Care* 9 (1997). See also the portrayal of the effects of shame in disease and age for a loving couple in Micheal Haneke's penetrating film *Amour*.

37. A. R. Hochschild, *The Managed Heart: The Commercialization of Human Feelings* (Berkeley: University of California Press, 1983).

38. M. Sommer, "Where the Education System and Women's Bodies Collide: The Social and Health Impact of Girls' Experiences of Menstruation and Schooling in Tanzania," *Journal of Adolescence* 33, no. 4 (2010).

39. T. Crofts and J. Fisher, "Menstrual Hygiene in Ugandan Schools: An Investigation of Low-Cost Sanitary Pads," https://dspace.lboro.ac.uk/dspace-jspui/handle/2134/9399, 2012.

40. B. O. Olatunji et al., "Multimodal Assessment of Disgust in Contamination-Related Obsessive-Compulsive Disorder," *Behaviour Research and Therapy* 45 (2007).

41. Rachman, "Fear of Contamination"; S. A. Rasmussen and J. L. Eisen, "Clinical and Epidemiologic Findings of Significance to Neuropharmacologic Trials in OCD," *Psychopharmacology Bulletin* 24, no. 3 (1988).

42. Tolin, Worhunsky, and Maltby, "Sympathetic Magic."

43. G. J. Rubin et al., "Public Perceptions, Anxiety and Behavioural Change in Relation to the Swine Flu Outbreak: A Cross-Sectional Telephone Survey," *British Medical Journal* 339 (2009); L. A. Page et al., "Using Electronic Patient Records to Assess the Impact of Swine Flu (Influenza H1N1) on Mental Health Patients," *Journal of Mental Health* 20, no. 1 (2011).

44. W. J. Magee et al., "Agoraphobia, Simple Phobia, and Social Phobia in the National Comorbidity Survey," *Archives of General Psychiatry* 53, no. 2 (1996).

45. P. Muris et al., "Disgust and Psychopathological Symptoms in a Nonclinical Sample," *Personality and Individual Differences* 29 (2000).

46. B. O. Olatunji and C. N. Sawchuk, "Disgust: Charcteristic Features, Social Manifestations and Clinical Implications," *Journal of Social and Clinical Psychology* 24, no. 7 (2005).

47. Davey et al., "Cross-Cultural Study of Animal Fears."

48. S. R. Woody, C. McLean, and T. Klassen, "Disgust as a Motivator of Avoidance of Spiders," *Journal of Anxiety Disorders* 19, no. 4 (2005).

49. Fleischman and Fessler, "Progesterone's Effects."

50. M. L. Phillips et al., "Disgust: The Forgotten Emotion of Psychiatry," *British Journal of Psychiatry* 172 (1998); G. C. L. Davey et al., "Disgust and Eating Disorders," *European Eating Disorders Review* 6, no. 3 (1998); N. A. Troop, J. L. Treasure, and L. Serpell, "A Further Exploration of Disgust in Eating Disorders," *European Eating Disorders Review* 10, no. 3 (2002); Muris et al., "Disgust and Psychopathological Symptoms"; B. Mayer et al., "Does Disgust Enhance Eating Disorder Symptoms?" *Eating Behaviors* 9, no. 1 (2008).

51. P. Rozin, M. Markwith, and C. Stoess, "Moralization and Becoming a Vegetarian: The Transformation of Preferences into Values and the Recruitment of Disgust," *Psychological Science* 8 (1997).

52. P. J. de Jong et al., "Disgust and Sexual Problems—Theoretical Conceptualization and Case Illustrations," *International Journal of Cognitive Therapy* 3, no. 1 (2010); C. Borg, P. J. de Jong, and W. W. Schultz, "Vaginismus and Dyspareunia: Automatic vs. Deliberate Disgust Responsivity," *Journal of Sexual Medicine* 7, no. 6 (2010).

53. De Jong et al., "Disgust and Sexual Problems."

54. S. Fry and U. Blumenbach, *Paperweight* (London: Arrow, 2004).

55. W. O. Monteiro et al., "Anorgasmia from Clomipramine in Obsessive Compulsive Disorder: A Controlled Trial," *British Journal of Psychiatry* 151 (1987).

56. R. Sprengelmeyer et al., "Disgust in Pre-Clinical Huntington's Disease: A Longitudinal Study," *Neuropsychologia* 44, no. 4 (2006).

57. C. May-Chahal and R. Antrobus, "Engaging Community Support in Safeguarding Adults from Self-Neglect," *British Journal of Social Work* 42 (2011).

58. T. Dalgleish and M. J. Power, "Emotion-Specific and Emotion-Non-Specific Components of Posttraumatic Stress Disorder (PTSD): Implications for a Taxonomy of Related Psychopathology," *Behaviour Research and Therapy* 42 (2004).

59. B. O. Olatunji et al., "Mental Pollution and PTSD Symptoms in Victims of Sexual Assault: A Preliminary Examination of the Mediating Role of Trauma-Related Cognitions," *Journal of Cognitive Psychotherapy* 22, no. 1 (2008).

60. De Jong et al., "Disgust and Sexual Problems."

61. P. J. de Jong and M. L. Peters, "Sex and the Sexual Dysfunctions: The Role of Disgust and Contamination Sensitivity," in *Disgust and Its Disorders: Theory, Assessment, and Treatment*, ed. B. O. Olatunji and D. McKay (Washington, DC: American Psychological Association, 2009).

62. Tolin, Worhunsky, and Maltby, "Sympathetic Magic."

63. De Jong et al., "Disgust and Sexual Problems."

64. Rubio-Godoy, Aunger, and Curtis, "Serotonin."

65. De Barra, "Attraction and Aversion."

66. M. Ridley, *The Origins of Virtue* (London: Viking, 1996); R. Wright, *The Moral Animal: Why We Are the Way We Are; The New Science of Evolutionary Psychology* (London: Vintage, 1995); J. Haidt, *The Happiness Hypothesis: Putting Ancient Wisdom and Philosophy to the Test of Modern Science* (London: Heinemann, 2007); Joyce, *Evolution of Morality.*

67. Jonathan Haidt, in his book *The Righteous Mind: Why Good People Are Divided by Politics and Religion*, points out that the Right focuses on the broad domain of morality in general, while the Left tends to be concerned much more closely with the rights of minorities. He suggests that it is time for the Left to reappropriate discussion of the big moral issues facing the majority about how it is best that we live.

68. Y. Inbar et al., "Disgust Sensitivity Predicts Intuitive Disapproval of Gays," *Emotion* 9, no. 3 (2009); K. Shanmugarajah et al., "The Role of Disgust Emotions in the Observer Response to Facial Disfigurement," *Body Image* 9 (2012); de Barra, "Attraction and Aversion."

69. D. Livingstone-Smith, *Less Than Human: Why We Demean, Enslave and Exterminate Others* (New York: St. Martin's Press, 2011).

70. V. Curtis, "The Dangers of Dirt: Household Hygiene and Health" (PhD diss., Wageningen Agricultural University, 1998). Steven Pinker has recently argued that one of the reasons for the decline in violence in recent history is the advance in hygiene. He suggests that it was easier to visit violence on someone who was repulsively dirty. S. Pinker, *The Better Angels of Our Nature: Why Violence Has Declined* (New York: Penguin Books, 2011).

71. K. E. Taylor, *Cruelty: Human Evil and the Human Brain* (Oxford: Oxford University Press, 2009).

72. J. Charteris-Black, "Britain as a Container: Immigration Metaphors in the 2005 Election Campaign," *Discourse and Society* 17, no. 5 (2006); S. I. Wilkinson, *Votes and Violence: Electoral Competition and Ethnic Riots in India* (Cambridge: Cambridge University Press, 2006).

73. Livingstone-Smith, *Less Than Human*.

74. L. Kass, "The Wisdom of Repugnance," *New Republic* 216, no. 22 (1997).

75. There is a major philosophical argument about whether one can determine what "ought to be" from what "is." I'm with Sam Harris, that we have an obligation to investigate the conditions that can enhance the flourishing of conscious beings. Harris, *Moral Landscape*. On moral progress, see also Steven Pinker's masterwork on the decline of violence: *The Better Angels of Our Nature* (London: Penguin, 2011).

76. P. Singer, *The Expanding Circle: Ethics and Sociobiology* (New York: Farrar, Straus and Giroux, 1981).

77. Livingstone-Smith, *Less Than Human*.

Epilogue

1. One of the reasons why psychology can't advance in the way that other sciences have is its continued failure to pin down and agree on its fundamental units of analysis—in brains or behavior. R. Aunger and V. Curtis, "Kinds of Behaviour," *Biology and Philosophy* 23, no. 3 (2008): 317–45.

2. Segerstråle, *Defenders of the Truth*; Laland and Brown, *Sense and Nonsense*.

3. There may of course be rare exceptions. Perhaps some extremophiles, animals in isolated niches where there is no competition from parasites, or animals with powerful physiological armor can do away with the need for behavioral avoidance of parasites.

4. Of course, the ways in which our living relative species "do" disgust will be quite varied, since many years have gone by for modifications to accumulate on each branch of the tree since the last common ancestor.

5. Borg, Lieberman, and Kiehl, "Infection, Incest, and Iniquity"; Lieberman, Tooby, and Cosmides, "Human Kin Detection"; J. M. Tybur, D. Lieberman, and V. Griskevicius, "Microbes, Mating, and Morality: Individual Differences in Three Functional Domains of Disgust," *Journal of Personality and Social Psychology* 97, no. 1 (2009).

6. Marsiglio, "Attitudes toward Homosexual Activity."

7. Isaksen, "Toward a Sociology of (Gendered) Disgust Images"; Ariely, *Predictably Irrational*; Shanmugarajah et al., "Role of Disgust Emotions."

8. B. Charlton, *Psychiatry and the Human Condition* (Oxford: Radcliffe Medical Press, 2000).

9. The approach is set out in more detail in Aunger and Curtis, "Kinds of Behaviour."

10. It's been proposed that men roll away after sex so as not get too high a dose of pair-bond-promoting oxytocin.

11. For example, components of the love system might include striving to be with the object of affection, sacrificing individual needs for the sake of the other, vigilance for rivals and chasing them away, and threatening serious consequences should a partner stray. Factor Analysis (and, eventually, high-resolution brain scanning) could help to distinguish the subcomponents of the love adaptive system.

BIBLIOGRAPHY

Aeby, G. S. "Trade-Offs for the Butterflyfish, *Chaetodon multicinctus*, When Feeding on Coral Prey Infected with Trematode Metacercariae." *Behavioral Ecology and Sociobiology* 52 (2002): 158–65.

Alio, A. P., L. Merrell, K. Roxburgh, H. B. Clayton, P. J. Marty, L. Bomboka, S. Traoré, and H. M. Salihu. "The Psychosocial Impact of Vesico-Vaginal Fistula in Niger." *Archives of Gynecology and Obstetrics* 284 (2011): 1–8.

Angyal, A. "Disgust and Related Aversions." *Journal of Abnormal and Social Psychology* 36 (1941): 393–412.

Arens, W. *The Man-Eating Myth: Anthropology and Anthropophagy.* Oxford: Oxford University Press, 1979.

Ariely, D. *Predictably Irrational: The Hidden Forces That Shape Our Decisions.* London: HarperCollins, 2008.

Aunger, R. "Are Food Avoidances Maladaptive in the Ituri Forest of Zaire?" *Journal of Anthropological Research* 50 (1994): 277–310.

Aunger, R., and V. Curtis. "Kinds of Behaviour." *Biology and Philosophy* 23, no. 3 (2008): 317–45.

Banbura, J., P. Perret, J. Blondel, A. Sauvages, M. J. Galan, and M. M. Lambrechts. "Sex Differences in Parental Care in a Corsican Blue Tit *Parus caeruleus* Population." *Ardea* 89, no. 3 (2001): 517–26.

Baumeister, R. F., K. D. Vohs, and C. Nathan DeWall. "How Emotion Shapes Behavior: Feedback, Anticipation, and Reflection, Rather Than Direct Causation." *Personality and Social Psychology Review* 11, no. 2 (2007): 167–203.

Behringer, D. C., M. J. Butler, and J. D. Shields. "Avoidance of Disease by Social Lobsters." *Nature* 441 (2006): 421.

Blair, R. H., and B. W. Tucker. "Nest Sanitation." *British Birds* 34 (1941): 206–15.

Borg, C., P. J. de Jong, and W. W. Schultz. "Vaginismus and Dyspareunia: Automatic vs. Deliberate Disgust Responsivity." *Journal of Sexual Medicine* 7, no. 6 (2010): 2149–57.

Borg, J., D. Lieberman, and K. A. Kiehl. "Infection, Incest, and Iniquity: Investigating the Neural Correlates of Disgust and Morality." *Journal of Cognitive Neuroscience* 20, no. 9 (2008): 1529–46.

Bourdieu, P. "Social Space and Symbolic Power." *Sociological Theory* 7, no. 1 (1989): 14–25.

Bowles, S., and H. Gintis. *A Cooperative Species: Human Reciprocity and Its Evolution*. Princeton, NJ: Princeton University Press, 2011.

Bradley, M., and K. Stow, eds. *Rome, Pollution and Propriety: Dirt, Disease and Hygiene in the Eternal City from Antiquity to Modernity*. Cambridge: Cambridge University Press, 2012.

Brashares, J. S., and P. Arcese. "Scent Marking in a Territorial African Antelope: II. The Economics of Marking with Faeces." *Animal Behaviour* 57, no. 1 (1999): 11–17.

Brian, A., W.-P. Schmidt, D. Rajaraman, K. Shankar, R. Aunger, and V. Curtis. "Superamma: A Cluster Randomized Trial of a Village-Level Intervention to Promote Handwashing with Soap in Rural Andhra Pradesh, India." Unpublished manuscript, 2013. Microsoft Word file.

British Heart Foundation. "Give Up before You Clog Up." Edited by UK Department of Health, 2003. Advertisement on YouTube, http://www.youtube.com/watch?v=cDAN7Oi62e0.

Brown, D. E. *Human Universals*. New York: McGraw-Hill, 1991.

Brown, G. E., D. P. Chivers, and R. J. F. Smith. "Localized Defecation by Pike: A Response to Labelling by Cyprinid Alarm Pheromone?" *Behavioral Ecology and Sociobiology* 36, no. 2 (1995): 105–10.

Carter, S. L. *Civility: Manners, Morals, and the Etiquette of Democracy*. New York: Basic Books, 1998.

Cashdan, E. "Adaptiveness of Food Learning and Food Aversions in Children." *Social Science Information* 37, no. 4 (1998): 613–32.

Chaisson, E. J. "Energy Rate Density as a Complexity Metric and Evolutionary Driver." *Complexity* 16, no. 3 (2011): 27–40.

Chapman, H. A., D. A. Kim, J. M. Susskind, and A. K. Anderson. "In Bad Taste: Evidence for the Oral Origins of Moral Disgust." *Science* 323, no. 5918 (2009): 1222.

Charlton, B. *Psychiatry and the Human Condition*. Oxford: Radcliffe Medical Press, 2000.

Charteris-Black, J. "Britain as a Container: Immigration Metaphors in the 2005 Election Campaign." *Discourse and Society* 17, no. 5 (2006): 563–81.

Clark, L., and J. R. Mason. "Use of Nest Material as Insecticidal and Anti-Pathogenic Agents by the European Starling." *Oecologia* 67, no. 2 (1985): 169–76.

Clayton, D. H. "Coevolution of Avian Grooming and Ectoparasite Avoidance." In *Bird-Parasite Interactions: Ecology, Evolution, and Behaviour*, edited by J. E. Loye and M. Zuk, 258–89. Oxford: Oxford University Press, 1991.

Cochran, G., and H. Harpending. *The 10,000 Year Explosion: How Civilization Accelerated Human Evolution*. New York: Basic Books, 2010.

Corbin, A. *The Foul and the Fragrant: Odor and the French Social Imagination*. Cambridge, MA: Harvard University Press, 1988.

Cremer, S., S. A. O. Armitage, and P. Schmid-Hempel. "Social Immunity." *Current Biology* 17, no. 16 (2007): R693–702.

Crofts, T., and J. Fisher. "Menstrual Hygiene in Ugandan Schools: An Investigation of Low-Cost Sanitary Pads." https://dspace .lboro.ac.uk/dspace-jspui/handle/2134/9399, 2012.

Curno, O., J. M. Behnke, A. G. McElligott, T. Reader, and C. J. Barnard. "Mothers Produce Less Aggressive Sons with Altered Immunity When There Is a Threat of Disease during Pregnancy." *Proceedings of the Royal Society of London B* 276, no. 1659 (2009): 1047–54.

Curtis, V. "Dirt, Disgust and Disease: A Natural History of Hygiene." *Journal of Epidemiology and Community Health* 61, no. 8 (2007): 660–64.

Curtis, V., R. Aunger, and T. Rabie. "Evidence That Disgust Evolved to Protect from Risk of Disease." *Proceedings of the Royal Society B* 271, supplement 4 (2004): S131–33.

Curtis, V., and A. Biran. "Dirt, Disgust, and Disease: Is Hygiene in Our Genes?" *Perspectives in Biology and Medicine* 44, no. 1 (2001): 17–31.

Curtis, V. A., A. Biran, K. Deverell, C. Hughes, K. Bellamy, and B. Drasar. "Hygiene in the Home: Relating Bugs to Behaviour." *Social Science and Medicine* 57, no. 4 (2003): 657–72.

Curtis, V., and S. Cairncross. "Effect of Washing Hands with Soap on Diarrhoea Risk in the Community: A Systematic Review." *Lancet Infectious Diseases* 3 (2003): 275–81.

Curtis, V., L. Danquah, and R. Aunger. "Planned, Motivated and Habitual Hygiene Behaviour: An Eleven Country Review." *Health Education Research* 24, no. 4 (2009): 655–73.

Curtis, V., M. de Barra, and R. Aunger. "Disgust as an Adaptive System for Disease Avoidance Behaviour." *Philosophical Transactions of the Royal Society B* 366 (2011): 389–401.

Curtis, V., B. Kanki, T. Mertens, T. Traore, I. Diallo, F. Tall, and S. Cousens. "Potties, Pits and Pipes: Explaining Hygiene Behaviour in Burkina Faso." *Social Science and Medicine* 41, no. 3 (1995): 383–93.

Curtis, V., W. Schmidt, S. Luby, R. Florez, O. Touré, and A. Biran. "Hygiene: New Hopes, New Horizons." *Lancet Infectious Diseases* 11, no. 4 (2011): 312–21.

Dahle, R. "Lukt, fukt og skam" (Bodily smells, wetness and shame). Paper presented at Norsk Selskap for Aldersforskning, Oslo, September 9, 1999.

Dalgleish, T., and M. J. Power. "Emotion-Specific and Emotion-Non-Specific Components of Posttraumatic Stress Disorder

(PTSD): Implications for a Taxonomy of Related Psychopathology." *Behaviour Research and Therapy* 42 (2004): 1069–88.

Damasio, A. R. *Descartes' Error: Emotion, Reason and the Human Brain*. New York: Putnam, 1994.

Darwin, C. *The Expression of the Emotions in Man and Animals*. 1872. Reprint. Chicago: University of Chicago Press, 1965.

Davey, G. C. L., G. Buckland, B. Tantow, and R. Dallos. "Disgust and Eating Disorders." *European Eating Disorders Review* 6, no. 3 (1998): 201–11.

Davey, G. C. L., McDonald, U. Hirisave, C. G. Prabhu, S. Iwawaki, I. J. Ching, H. Merckelbach, P. J. de Jong, P. W. L. Leung, and B. C. Reimann. "A Cross-Cultural Study of Animal Fears." *Behaviour Research and Therapy* 36 (1998): 735–50.

David, B., and B. O. Olatunji. "The Effect of Disgust Conditioning and Disgust Sensitivity on Appraisals of Moral Transgressions." *Personality and Individual Differences* 50, no. 7 (2011): 1142–46.

Dawkins, R. *The Selfish Gene*. New York: Oxford University Press, 1976.

———. *Unweaving the Rainbow*. London: Allen Lane, 1998.

De Barra, M. "Attraction and Aversion: Pathogen Avoidance Strategies in the UK and Bangladesh." PhD diss., London School of Hygiene and Tropical Medicine, 2011.

Decaestecker, E., L. De Meester, and D. Ebert. "In Deep Trouble: Habitat Selection Constrained by Multiple Enemies in Zooplankton." *Proceedings of the National Academy of Sciences USA* 99, no. 8 (2002): 5481–85.

De Jong, P. J., and M. L. Peters. "Sex and the Sexual Dysfunctions: The Role of Disgust and Contamination Sensitivity." In *Disgust and Its Disorders: Theory, Assessment, and Treatment*, ed. B. O. Olatunji and D. McKay, 253–70. Washington, DC: American Psychological Association, 2009.

De Jong, P. J., J. van Lankveld, H. J. Elgersma, and C. Borg. "Disgust and Sexual Problems—Theoretical Conceptualization and

Case Illustrations." *International Journal of Cognitive Therapy* 3, no. 1 (2010): 23–39.

Dethier, V. G. *The World of the Tent-Makers: A Natural History of the Eastern Tent Caterpillar.* Amherst: University of Massachusetts Press, 1980.

Diamond, J. M. "Archaeology: Talk of Cannibalism." *Nature* 407, no. 6800 (2000): 25–26.

Diehl-Fleig, E., and M. E. Lucchese. "Reações comportamentais de operárias de *Acromyrmex striatus* (Hymenoptera, Formicidae) na presença de fungos entomopatogênicos." *Revista Brasileira de Entomologia* 35 (1991): 101–7.

Douglas, M. *Purity and Danger: An Analysis of Concepts of Pollution and Taboo.* London: Routledge, 2003.

Duncan, L. A., and M. Schaller. "Prejudicial Attitudes toward Older Adults May Be Exaggerated When People Feel Vulnerable to Infectious Disease: Evidence and Implications." *Analyses of Social Issues and Public Policy* 9, no. 1 (2009): 97–115.

Durham, W. H. *Coevolution: Genes, Culture and Human Diversity.* Stanford, CA: Stanford University Press, 1991.

Earle, S. "Factors Affecting the Initiation of Breastfeeding: Implications for Breastfeeding Promotion." *Health Promotion International* 17, no. 3 (2002): 205.

Ehrlich, P. R., D. S. Dobkin, and D. Wheye. "The Adaptive Significance of Anting." *Auk* 103, no. 4 (1986): 835.

Eisner T., E. van Tassell, and J. E. Carrel. "Defensive Use of a 'Fecal Shield' by a Beetle Larva." *Science* 158 (1967): 1471–73.

Elias, N. *The Civilizing Process.* 1939. Translated by E. Jephcott. Malden, MA: Blackwell, 2000.

Emerson, P. M., S. Cairncross, R. L. Bailey, and D. C. W. Mabey. "A Review of the Evidence for the 'F' and 'E' Components of the SAFE Strategy for Trachoma Control." *Tropical Medicine and International Health* 5, no. 8 (2000): 515–27.

Eskine, K. J., N. A. Kacinik, and J. J. Prinz. "A Bad Taste in the Mouth." *Psychological Science* 22, no. 3 (2011): 295.

Evans, D. B., and S. F. Hurley. "The Application of Economic Evaluation Techniques in the Health Sector: The State of the Art." *Journal of International Development* 7, no. 3 (1995): 503–24.

Fayard, J. V., A. K. Bassi, D. M. Bernstein, and B. W. Roberts. "Is Cleanliness Next to Godliness? Dispelling Old Wives' Tales: Failure to Replicate Zhong and Liljenquist (2006)." *Journal of Articles in Support of the Null Hypothesis* 6, no. 2 (2009): 21–28.

Fessler, D. M. T. "Shame in Two Cultures: Implications for Evolutionary Approaches." *Journal of Cognition and Culture* 4, no. 2 (2004): 207–62.

Fessler, D. M. T., S. J. Eng, and C. D. Navarrete. "Elevated Disgust Sensitivity in the First Trimester of Pregnancy: Evidence Supporting the Compensatory Prophylaxis Hypothesis." *Evolution and Human Behavior* 26 (2005): 344–51.

Flaxman, S. M., and P. W. Sherman. "Morning Sickness: A Mechanism for Protecting Mother and Embryo." *Quarterly Review of Biology* 75, no. 2 (2000): 113–48.

Fleischman, D. S., M. de Barra, G. Webster, R. Aunger, G. Judah, and V. Curtis. "Sensor Recorded Changes in Rates of Hand Washing with Soap in Response to the Media Reports of the H1N1 Pandemic in Britain." *BMJ Open* 1 (2011).

Fleischman, D. S., and D. M. T. Fessler. "Progesterone's Effects on the Psychology of Disease Avoidance: Support for the Compensatory Behavioral Prophylaxis Hypothesis." *Hormones and Behavior* 59, no. 2 (2011): 271–75.

Folstad, I., O. Halvorsen, and A. C. Nilssen. "Parasite Avoidance: The Cause of Post-Calving Migrations in *Rangifer*?" *Canadian Journal of Zoology* 69, no. 9 (1991): 2423–29.

Franks, N. R., J. W. Hooper, C. Webb, and A. Dornhaus. "Tomb Evaders: House-Hunting Hygiene in Ants." *Biology Letters* 1 (2005): 190–92.

Freeland, W. J. "Pathogens and the Evolution of Primate Sociality." *Biotropica* 8, no. 1 (1976): 12–24.

Fretwell, P. T., and P. N. Trathan. "Penguins from Space: Faecal

Stains Reveal the Location of Emperor Penguin Colonies." *Global Ecology and Biogeography* 18, no. 5 (2009): 543–52.

Freud, S., J. Strachey, and A. Richards. *On Sexuality: Three Essays on the Theory of Sexuality and Other Works (1905)*. London: Penguin Books, 1977.

Fry, S., and U. Blumenbach. *Paperweight*. London: Arrow, 2004.

Fumagalli, M., M. Sironi, U. Pozzoli, A. Ferrer-Admettla, L. Pattini, and R. Nielsen. "Signatures of Environmental Genetic Adaptation Pinpoint Pathogens as the Main Selective Pressure through Human Evolution." *PLOS Genetics* 7, no. 11 (2011): e1002355. doi:10.1371/journal.pgen.1002355.

Fung, I. C.-H., and S. Cairncross. "Effectiveness of Handwashing in Preventing SARS: A Review." *Tropical Medicine and International Health* 11, no. 11 (2006): 1749–58.

Gajdusek, D. C. "Unconventional Viruses and the Origin and Disappearance of Kuru." In *Nobel Lectures in Physiology or Medicine 1971–1980*, ed. J. Lindsten, 305–54. Stockholm: Karolinska Institute, 1977.

Garcia, J., D. Forthman-Quick, and B. White. "Conditioned Disgust and Fear from Mollusk to Monkey." In *Primary Neural Substrates of Learning and Behavioral Change*, ed. D. L. Alkon and J. Farley, 47–61. New York: Cambridge University Press, 1984.

Gehlbach, F. R., and R. S. Baldridge. "Live Blind Snakes (*Leptotyphlops dulcis*) in Eastern Screech Owl (*Otus asio*) Nests: A Novel Commensalism." *Oecologia* 71, no. 4 (1987): 560–63.

Gilbert, G., director. *Jamie's School Dinners*. 48 minutes. Channel 4 (UK), 2005.

Giner-Sorolla, R., and P. Espinosa. "Social Cuing of Guilt by Anger and of Shame by Disgust." *Psychological Science* 22, no. 1 (2011): 49–53.

Gottman, J. M. *What Predicts Divorce? The Relationship between Marital Processes and Marital Outcomes*. Hillsdale, NJ: Lawrence Erlbaum, 1994.

Greenland, K., S. Cairncross, O. Cumming, and V. Curtis. "Editorial: Can We Afford to Overlook Hand Hygiene Again?" *Tropical Medicine and International Health* 18, no. 3 (2013): 246–49.

Gumert, M. D. "Payment for Sex in a Macaque Mating Market" *Animal Behaviour* 74, no. 6 (2007): 1655–67.

Gunn, A., and R. J. Irvine. "Subclinical Parasitism and Ruminant Foraging Strategies: A Review." *Wildlife Society Bulletin* 31, no. 1 (2003): 117–26.

Haidt, J. *The Happiness Hypothesis: Putting Ancient Wisdom and Philosophy to the Test of Modern Science*. London: Heinemann, 2007.

———. *The Righteous Mind: Why Good People Are Divided by Politics and Religion*. London: Penguin Books, 2010.

Haidt, J., C. McCauley, and P. Rozin. "Individual Differences in Sensitivity to Disgust: A Scale Sampling Seven Domains of Disgust Elicitors." *Personality and Individual Differences* 16, no. 5 (1994): 701–13.

Hamilton, W. "The Genetical Evolution of Social Behaviour." *Journal of Theoretical Biology* 7 (1964): 1–52.

Hammond, D., G. T. Fong, P. W. McDonald, S. Brown, and R. Cameron. "Graphic Canadian Cigarette Warning Labels and Adverse Outcomes: Evidence from Canadian Smokers." *American Journal of Public Health* 94, no. 8 (2004): 1442–45.

Harris, S. *The Moral Landscape: How Science Can Determine Human Values*. New York: Free Press, 2011.

Hart, A. G., and F. L. W. Ratnieks. "Task Partitioning, Division of Labour and Nest Compartmentalisation Collectively Isolate Hazardous Waste in the Leafcutting Ant *Atta cephalotes*." *Behavioral Ecology and Sociobiology* 49, no. 5 (2001): 387–92.

Hart, B. L. "Behavioural Defences in Animals against Pathogens and Parasites: Parallels with the Pillars of Medicine in Humans." *Philosophical Transactions of the Royal Society B* 366, no. 1583 (2011): 3406–17.

Hart, B. L., E. Korinek, and P. Brennan. "Postcopulatory Genital Grooming in Male Rats: Prevention of Sexually Transmitted Infections." *Physiology and Behavior* 41, no. 4 (1987): 321–25.

Haselton, M. G., and D. Nettle. "The Paranoid Optimist: An Integrative Evolutionary Model of Cognitive Biases." *Personality and Social Psychology Review* 10, no. 1 (2006): 47–66.

Hennig, J., P. Pössel, and P. Netter. "Sensitivity to Disgust as an Indicator of Neuroticism: A Psychobiological Approach." *Personality and Individual Differences* 20, no. 5 (1996): 589–96.

Henrich, J. "The Evolution of Costly Displays, Cooperation and Religion: Credibility Enhancing Displays and Their Implications for Cultural Evolution." *Evolution and Human Behavior* 30, no. 4 (2009): 244–60.

Herz, R. *That's Disgusting: Unraveling the Mysteries of Repulsion.* New York: W. W. Norton and Company, 2012.

Heymann, D., ed. *Control of Communicable Diseases Manual.* 19th ed. Washington, DC: American Public Health Association, 2008.

Hochschild, A. R. *The Managed Heart: The Commercialization of Human Feelings.* Berkeley: University of California Press, 1983.

Hoefling, A., K. U. Likowski, R. Deutsch, M. Häfner, B. Seibt, A. Mühlberger, P. Weyers, and F. Strack. "When Hunger Finds No Fault with Moldy Corn: Food Deprivation Reduces Food-Related Disgust." *Emotion* 9, no. 1 (2009): 50–58.

Hölldobbler, B., and E. O. Wilson. *The Ants.* Cambridge, MA: Harvard University Press, 1990.

Hosoi, S. A., and S. I. Rothstein. "Nest Desertion and Cowbird Parasitism: Evidence for Evolved Responses and Evolutionary Lag." *Animal Behaviour* 59, no. 4 (2000): 823–40.

Huang, D. B., and J. Zhou. "Effect of Intensive Handwashing in the Prevention of Diarrhoeal Illness among Patients with AIDS: A Randomized Controlled Study." *Journal of Medical Microbiology* 56, no. 5 (2007): 659–63.

Humphrey, J. H. "Child Undernutrition, Tropical Enteropathy, Toilets, and Handwashing." *Lancet* 374 (2009): 1032–35.

Hurtrez-Boussès, S., F. Renaud, J. Blondel, P. Perret, and M. J. Galan. "Effects of Ectoparasites of Young on Parents' Behaviour in a Mediterranean Population of Blue Tits." *Journal of Avian Biology* 31, no. 2 (2000): 266–69.

Hutchings, M. R., I. Kyriazakis, T. G. Papachristou, I. J. Gordon, and F. Jackson. "The Herbivores' Dilemma: Trade-Offs between Nutrition and Parasitism in Foraging Decisions." *Oecologia* 124 (2000): 242–51.

Inbar, Y., D. A. Pizarro, J. Knobe, and P. Bloom. "Disgust Sensitivity Predicts Intuitive Disapproval of Gays." *Emotion* 9, no. 3 (2009): 435–39

Isaksen, L. W. "Toward a Sociology of (Gendered) Disgust Images of Bodily Decay and the Social Organization of Care Work." *Journal of Family Issues* 23, no. 7 (2002): 791–811.

Jansen, G. C. M. , A. O. Koloski-Ostrow, and E. M. Moormann, eds. *Roman Toilets: Their Archaeology and Cultural History.* Leuven, Belgium: Peeters Publishers, 2011.

Jefferson, T., R. Foxlee, C. Del Mar, L. Dooley, E. Ferroni, B. Hewak, A. Prabhala, S. Nair, and A. Rivetti. "Physical Interventions to Interrupt or Reduce the Spread of Respiratory Viruses: Systematic Review." *British Medical Journal* 336 (2008): 77–80.

Joyce, R. *The Evolution of Morality.* Cambridge, MA: MIT Press, 2006.

Judah, G., R. Aunger, W. P. Schmidt, S. Michie, S. Granger, and V. Curtis. "Experimental Pretesting of Hand-Washing Interventions in a Natural Setting." *American Journal of Public Health* 99, no. S2 (2009): S405–11.

Karvonen, A., O. Seppala, and E. T. Valtonen. "Parasite Resistance and Avoidance Behaviour in Preventing Eye Fluke Infections in Fish." *Parasitology* 129, no. 2 (2004): 159–64.

Kass, L. "The Wisdom of Repugnance." *New Republic* 216, no. 22 (1997): 17–26.

Kavaliers, M., and D. D. Colwell. "Discrimination by Female Mice between the Odours of Parasitized and Non-Parasitized Males." *Proceedings of the Royal Society of London B* 261 (1995): 31–35.

Keisecker, J., D. Skelly, K. Beard, and E. Preisser. "Behavioral Reduction of Infection Risk." *Proceedings of the National Academy of Sciences USA* 96 (1999): 9165–68.

Kelly, D. *Yuck! The Nature and Moral Significance of Disgust.* Cambridge, MA: MIT Press, 2011.

Kolnai, A. *On Disgust.* 1927. Translated by E. Kolnai. Peru, IL: Carus, 2004.

Konrad, M., M. L. Vyleta, F. J. Theis, and S. Cremer. "Social Transfer of Pathogenic Fungus Promotes Active Immunisation in Ant Colonies." *PLOS Biology* 10, no. 4 (2012): e1001300.

Korsmeyer, C. *Savoring Disgust: The Foul and the Fair in Aesthetics.* Oxford: Oxford University Press, 2011.

Krause, J., and J.-G. J. Godin. "Influence of Parasitism on the Shoaling Behaviour of Banded Killifish, *Fundulus diaphanus.*" *Canadian Journal of Zoology* 72 (1994): 1775–79.

Krone, R., R. Bshary, M. Paster, M. Eisinger, P. van Treeck, and H. Schuhmacher. "Defecation Behaviour of the Lined Bristletooth Surgeonfish *Ctenochaetus striatus* (Acanthuridae)." *Coral Reefs* 27, no. 3 (2008): 619–22.

Kuris, A. M., R. F. Hechinger, J. C. Shaw, K. L. Whitney, L. Aguirre-Macedo, C. A. Boch, A. P. Dobson, et al. "Ecosystem Energetic Implications of Parasite and Free-Living Biomass in Three Estuaries." *Nature* 454, no. 7203 (2008): 515–18.

Kurzban, R., and M. R. Leary. "Evolutionary Origins of Stigmatization: The Functions of Social Exclusion." *Psychological Bulletin* 127, no. 2 (2001): 187–208.

Laland, K. N., N. Atton, and M. M. Webster. "From Fish to Fashion: Experimental and Theoretical Insights into the Evolution of Culture." *Philosophical Transactions of the Royal Society B* 366, no. 1567 (2011): 958–68.

Laland, K. N., and G. R. Brown. *Sense and Nonsense: Evolutionary Perspectives on Human Behaviour.* Oxford: Oxford University Press, 2011.

Lamont, M. *Money, Morals and Manners: The Culture of the French and American Upper-Middle Class*. Chicago: University of Chicago Press, 1992.

Lévi-Strauss, C. *The Origin of Table Manners*. Vol. 3 of *Mythologiques*. Chicago: University of Chicago Press, 1990.

Lieberman, D., J. Tooby, and L. Cosmides. "The Architecture of Human Kin Detection." *Nature* 445, no. 7129 (2007): 727–31.

Liljenquist, K., C. B. Zhong, and A. D. Galinsky. "The Smell of Virtue: Clean Scents Promote Reciprocity and Charity." *Psychological Science* 21, no. 3 (2010): 381–83.

Lindenbaum, S. "Understanding Kuru: The Contribution of Anthropology and Medicine." *Philosophical Transactions of the Royal Society B* 363, no. 1510 (2008): 3715–20.

Liu, L., H. L. Johnson, S. Cousens, J. Perin, S. Scott, J. E. Lawn, I. Rudan, H. Campbell, R. Cibulskis, and M. Li. "Global, Regional, and National Causes of Child Mortality: An Updated Systematic Analysis for 2010 with Time Trends since 2000." *Lancet* 379 (2012): 2151–61.

Livingstone-Smith, D. *Less Than Human: Why We Demean, Enslave and Exterminate Others*. New York: St. Martin's Press, 2011.

Loehle, C. "Social Barriers to Pathogen Transmission in Wild Animal Populations." *Ecology* 76 (1995): 326–35.

Longino, J. T. "True Anting by the Capuchin, *Cebus capucinus*." *Primates* 25, no. 2 (1984): 243–45.

Magee, W. J., W. W. Eaton, H. U. Wittchen, K. A. McGonagle, and R. C. Kessler. "Agoraphobia, Simple Phobia, and Social Phobia in the National Comorbidity Survey." *Archives of General Psychiatry* 53, no. 2 (1996): 159–68.

Magin, P., J. Adams, G. Heading, D. Pond, and W. Smith. "Psychological Sequelae of Acne Vulgaris: Results of a Qualitative Study." *Canadian Family Physician* 52, no. 8 (2006): 978.

Mameli, M., and P. Bateson. "Innateness and the Sciences." *Biology and Philosophy* 21, no. 2 (2006): 155–88.

Marcus, G. F. *The Birth of the Mind: How a Tiny Number of Genes Creates the Complexities of Human Thought.* New York: Basic Books, 2004.

Marlar, R. A., B. L. Leonard, B. R. Billman, P. M. Lambert, and J. E. Marlar. "Biochemical Evidence of Cannibalism at a Prehistoric Puebloan Site in Southwestern Colorado." *Nature* 407, no. 6800 (2000): 74–78.

Marsiglio, W. "Attitudes toward Homosexual Activity and Gays as Friends: A National Survey of Heterosexual 15!to 19YearOld Males." *Journal of Sex Research* 30, no. 1 (1993): 12–17.

Mathew, S., and R. Boyd. "Punishment Sustains Large-Scale Cooperation in Prestate Warfare." *Proceedings of the National Academy of Sciences USA* 108, no. 28 (2011): 11375–80

May-Chahal, C., and R. Antrobus. "Engaging Community Support in Safeguarding Adults from Self-Neglect." *British Journal of Social Work* 42 (2011): 1478–94.

Mayer, B., A. E. R. Bos, P. Muris, J. Huijding, and M. Vlielander. "Does Disgust Enhance Eating Disorder Symptoms?" *Eating Behaviors* 9, no. 1 (2008): 124–27.

Mayfield, H. "Nesting Success Calculated from Exposure." *Wilson Bulletin* 73 (1961): 255–61.

McGinn, C. *The Meaning of Disgust.* Oxford: Oxford University Press, 2011.

Mennerat, A., P. Mirleau, J. Blondel, P. Perret, M. M. Lambrechts, and P. Heeb. "Aromatic Plants in Nests of the Blue Tit *Cyanistes caeruleus* Protect Chicks from Bacteria." *Oecologia* 161, no. 4 (2009): 849–55.

Menninghaus, W. *Disgust: The Theory and History of a Strong Sensation.* New York: State University of New York Press, 2003.

Meyer-Rochow, V. B., and J. Gal. "Pressures Produced When Penguins Pooh: Calculations on Avian Defaecation." *Polar Biology* 27 (2003): 56–58.

Miller, S. B. *Disgust: The Gatekeeper Emotion.* Hillsdale, NJ: Analytic Press, 2004.

Miller, W. I. *The Anatomy of Disgust*. Cambridge, MA: Harvard University Press, 1997.

Mineka, S., and M. Cook. "Mechanisms Involved in the Observational Conditioning of Fear." *Journal of Experimental Psychology, General* 122 (1993): 23–38.

Mithen, S. *The Prehistory of the Mind: A Search for the Origins of Art, Religion and Science*. London: Thames and Hudson, 1996.

Moll, J., R. de Oliveira-Souza et al. "The Moral Affiliations of Disgust: A Functional MRI Study." *Cognitive and Behavioral Neurology* 18, no. 1 (2005): 68–78.

Monteiro W. O., H. F. Noshirvani, I. M. Marks, and P. T. Lelliott. "Anorgasmia from Clomipramine in Obsessive Compulsive Disorder: A Controlled Trial." *British Journal of Psychiatry* 151 (1987): 107–12.

Montevecchi, W. A. "Eggshell Removal by Laughing Gulls." *Bird-Banding* 47, no. 2 (1976): 129–35.

Mooring, M .S., A. A. McKenzie, and B. L. Hart. "Grooming in Impala: Role of Oral Grooming in Removal of Ticks and Effects of Ticks in Increasing Grooming Rate." *Physiology and Behavior* 59 (1996): 965–71.

Muris, P., H. Merckelbach, S. Nederkoorn, E. Rassin, I. Candel, and R. Horselenberg. "Disgust and Psychopathological Symptoms in a Nonclinical Sample." *Personality and Individual Differences* 29 (2000): 1163–67.

Nesse, R. M. . "Natural Selection and the Regulation of Defenses: A Signal Detection Analysis of the Smoke Detector Principle." *Evolution and Human Behavior* 26 (2005): 88–105.

Nesse, R. M., and G. C. Williams. *Evolution and Healing*. London: Weidenfeld and Nicolson, 1995.

Nettle, D. "The Evolution of Personality Variation in Humans and Other Animals." *American Psychologist* 61, no. 6 (2006): 622.

———. *Personality: What Makes You the Way You Are*. Oxford: Oxford University Press, 2007.

Nichols, S. "On the Genealogy of Norms: A Case for the Role of

Emotion in Cultural Evolution." *Philosophy of Science* 69 (2002): 234–55.

Norris, K. "A Trade-Off between Energy Intake and Exposure to Parasites in Oystercatchers Feeding on a Bivalve Mollusc." *Proceedings of the Royal Society of London B* 266, no. 1429 (1999): 1703–9.

Nowack, M., and R. Highfield. *Supercooperators: Evolution, Altruism and Human Behaviour; or, Why We Need Each Other to Succeed.* New York: Free Press, 2011.

Nowak, M. A., and K. Sigmund. "Evolution of Indirect Reciprocity." *Nature* 437, no. 7063 (2005): 1291–98.

Nussbaum, M. *Hiding from Humanity: Disgust, Shame and the Law.* Princeton, NJ: Princeton University Press, 2004.

Oaten, M., R. J. Stevenson, and T. I. Case. "Disease Avoidance as a Functional Basis for Stigmatization." *Philosophical Transactions of the Royal Society B* 366, no. 1583 (2011): 3433–52.

O'Hara, S. J., and P. C. Lee. "High Frequency of Post-Coital Penis Cleaning in Budongo Chimpanzees." *Folia Primatologica* 77, no. 5 (2006): 353–58.

Olatunji, B. O., L. S. Elwood, N. L. Williams, and J. M. Lohr. "Mental Pollution and PTSD Symptoms in Victims of Sexual Assault: A Preliminary Examination of the Mediating Role of Trauma-Related Cognitions." *Journal of Cognitive Psychotherapy* 22, no. 1 (2008): 37–47.

Olatunji, B. O., J. P. Forsyth, and A. Cherian. "Evaluative Differential Conditioning of Disgust: A Sticky Form of Relational Learning That Is Resistant to Extinction." *Journal of Anxiety Disorders* 21 (2007): 820–34.

Olatunji, B. O., J. M. Lohr, C. N. Sawchuk, and D. F. Tolin. "Multimodal Assessment of Disgust in Contamination-Related Obsessive-Compulsive Disorder." *Behaviour Research and Therapy* 45 (2007): 263–76.

Olatunji, B. O., and C. N. Sawchuk. "Disgust: Charcteristic Features, Social Manifestations and Clinical Implications." *Journal of Social and Clinical Psychology* 24, no. 7 (2005): 932–62.

Oppliger, A., H. Richner, and P. Christie. "Effect of an Ectoparasite on Lay Date, Nest Site Choice, Desertion and Hatching Success in the Great Tit (*Parus major*)." *Behavioural Ecology* 5, no. 2 (1994): 130–34.

Packer, A. J., and P. Espeland. *How Rude! The Teenagers' Guide to Good Manners, Proper Behavior, and Not Grossing People Out*. Minneapolis, MN: Free Spirit Publishing, 1997.

Page, L. A., S. Seetharaman, I. Suhail, S. Wessely, J. Pereira, and G. J. Rubin. "Using Electronic Patient Records to Assess the Impact of Swine Flu (Influenza H1N1) on Mental Health Patients." *Journal of Mental Health* 20, no. 1 (2011): 60–69.

Pagel, M., and W. Bodmer. "A Naked Ape Would Have Fewer Parasites." *Biology Letters* 270, no. S1 (2003): 117–19.

Panksepp, J. *Affective Neuroscience*. Oxford: Oxford University Press, 1998.

Park, J. H., J. Faulkner, and M. Schaller. "Evolved Disease-Avoidance Processes and Contemporary Anti-Social Behavior: Prejudicial Attitudes and Avoidance of People with Physical Disabilities." *Journal of Nonverbal Behavioral and Brain Sciences* 27 (2003): 65–87.

Park, J. H., M. Schaller, and C. S. Crandall. "Pathogen-Avoidance Mechanisms and the Stigmatization of Obese People." *Evolution and Human Behavior* 28, no. 6 (2007): 410–14.

Pepper, G. V., and S. C. Roberts. "Rates of Nausea and Vomiting in Pregnancy and Dietary Characteristics across Populations." *Proceedings of the Royal Society B* 273, no. 1601 (2006): 2675–79.

Petit, C., M. Hossaert-McKey, P. Perret, J. Blondel, and M. M. Lambrechts. "Blue Tits Use Selected Plants and Olfaction to Maintain an Aromatic Environment for Nestlings." *Ecology Letters* 5 (2002): 585–89.

Pfennig, D. W., S. G. Ho, and E. A. Hoffman. "Pathogen Transmission as a Selective Force against Cannibalism." *Animal Behaviour* 55, no. 5 (1998): 1255–61.

Pfennig, D. W., M. L. G. Loeb, and J. P. Collins. "Pathogens as a Factor Limiting the Spread of Cannibalism in Tiger Salamanders." *Oecologia* 88 (1991): 161–66.

Phillips, M. L., C. Senior, T. Fahy, and A. S. David. "Disgust: The Forgotten Emotion of Psychiatry." *British Journal of Psychiatry* 172 (1998): 373–75.

Pie, M. R., R. B. Rosengaus, and J. F. A. Traniello. "Nest Architecture, Activity Pattern, Worker Density and the Dynamics of Disease Transmission in Social Insects." *Journal of Theoretical Biology* 226, no. 1 (2004): 45–51.

Pinker, S. *The Better Angels of Our Nature: Why Violence Has Declined.* London: Penguin, 2011.

———. "The False Allure of Group Selection." *Edge.* http://edge.org/conversation/the-false-allure-of-group-selection, 2012.

———. *How the Mind Works.* London: Penguin, 1998.

Polunin, N. V. C., and I. Koike. "Temporal Focusing of Nitrogen Release by a Periodically Feeding Herbivorous Reef Fish." *Journal of Experimental Marine Biology and Ecology* 111, no. 3 (1987): 285–96.

Rabie, T., and V. Curtis. "Handwashing and Risk of Respiratory Infections: A Quantitative Systematic Review." *Tropical Medicine and International Health* 11, no. 3 (2006): 269–78.

Rachman, S. "Fear of Contamination." *Behaviour Research and Therapy* 42 (2004): 1227–55.

Rasmussen, S. A., and J. L. Eisen. "Clinical and Epidemiologic Findings of Significance to Neuropharmacologic Trials in OCD." *Psychopharmacology Bulletin* 24, no. 3 (1988): 466–70.

Rawdon Wilson, R. *The Hydra's Tale: Imagining Disgust.* Edmonton: University of Alberta Press, 2002.

Reinhart, A. K. "Impurity/No Danger." *History of Religions* 30, no. 1 (1990): 1–24.

Richerson, P. J., and R. Boyd. *Not by Genes Alone: How Culture Trans-*

formed Human Evolution. Chicago: University of Chicago Press, 2005.

Ridley, M. *The Origins of Virtue.* London: Viking, 1996.

———. *The Rational Optimist: How Prosperity Evolves.* London: Fourth Estate, 2010.

Royzman, E., and R. Kurzban. "Minding the Metaphor: The Elusive Character of Moral Disgust." *Emotion Review* 3, no. 3 (2011): 269–71.

Rozin, P., J. Haidt, and C. R. McCauley. "Disgust." In *Handbook of Emotion,* ed. M. Lewis, J. M. Haviland-Jones, and L. F. Barrett, 757–76. New York: Guilford Press, 2008.

Rozin, P., J. Haidt, C. McCauley, L. Dunlop, and M. Ashmore. "Individual Differences in Disgust Sensitivity: Comparisons and Evaluations of Paper-and-Pencil versus Behavioral Measures." *Journal of Research in Personality* 33, no. 3 (1999): 330–51.

Rozin, P., M. Markwith, and C. Stoess. "Moralization and Becoming a Vegetarian: The Transformation of Preferences into Values and the Recruitment of Disgust." *Psychological Science* 8 (1997): 67–73.

Rubin, G. J., R. Amlôt, L. Page, and S. Wessely. "Public Perceptions, Anxiety and Behavioural Change in Relation to the Swine Flu Outbreak: A Cross-Sectional Telephone Survey." *British Medical Journal* 339 (2009): b2651.

Rubin, G. J., H. W. W. Potts, and S. Michie. "The Impact of Communications about Swine Flu (Influenza a H1N1v) on Public Responses to the Outbreak: Results from 36 National Telephone Surveys in the UK." *Health Technology Assessment* 14, no. 34 (2010): 183–266.

Rubio-Godoy, M., R. Aunger, and V. Curtis. "Serotonin: A Link between Disgust and Immunity?" *Medical Hypotheses* 68, no. 1 (2007): 61–66.

Sanfey, A. G., J. K. Rilling, J. A. Aronson, L. E. Nystrom, and J. D. Cohen. "The Neural Basis of Economic Decision-Making in the Ultimatum Game." *Science* 300, no. 5626 (2003): 1755–58.

Sato, Y., Y. Saito, and T. Sakagami. "Rules for Nest Sanitation in a Social Spider Mite, *Schizotetranychus miscanthi* Saito (Acari: Tetranychidae)." *Ethology* 109, no. 9 (2003): 713–24.

Savigny, J. B. H., and A. Corréard. *Narrative of a Voyage to Senegal in 1816: Undertaken by Order of the French Government, Comprising an Account of the Shipwreck of the Medusa, the Sufferings of the Crew, and the Various Occurrences on Board the Raft, in the Desert of Zaara, at St. Louis, and at the Camp of Daccard. To Which Are Subjoined Observations Respecting the Agriculture of the Western Coast of Africa, from Cape Blanco to the Mouth of the Gambia.* London: H. Colburn, 1818.

Schaller, M., G. E. Miller, W. M. Gervais, S. Yager, and E. Chen. "Mere Visual Perception of Other People's Disease Symptoms Facilitates a More Aggressive Immune Response." *Psychological Science* 21, no. 5 (2010): 649–52.

Schmid-Hempel, P. *Parasites in Social Insects.* Princeton, NJ: Princeton University Press, 1998.

Schmidt, C. W. "The Yuck Factor: Where Disgust Meets Discovery." *Environmental Health Perspectives* 116, no. 12 (2008): A524–27.

Schnall, S., J. Benton, and S. Harvey. "With a Clean Conscience." *Psychological Science* 19, no. 12 (2008): 1219–22.

Schnall, S., J. Haidt, G. L. Clore, and A. H. Jordan. "Disgust as Embodied Moral Judgment." *Personality and Social Psychology Bulletin* 34, no. 8 (2008): 1096–1109.

Schulenburg, H., and S. Muller. "Natural Variation in the Response of *Caenorhabditis elegans* towards *Bacillus thuringiensis*." *Parasitology* 128 (2004): 433–43.

Scott, B. E., W. P. Schmidt, R. Aunger, N. Garbrah-Aidoo, and R. Animashaun. "Marketing Hygiene Behaviours: The Impact of Different Communications Channels on Reported Handwashing Behaviour of Women in Ghana." *Health Education Research* 22, no. 4 (2007): 225–33.

Segerstråle, U. C. O. *Defenders of the Truth: The Sociobiology Debate.* Oxford: Oxford University Press, 2001.

Seligman, M. E. P. "Phobias and Preparedness." *Behavior Therapy* 2, no. 3 (1971): 307–20.

Shanmugarajah, K., S. Gaind, A. Clarke, and P. E. M. Butler. "The Role of Disgust Emotions in the Observer Response to Facial Disfigurement." *Body Image* 9 (2012): 455–61.

Shettleworth, S. J. "Modularity, Comparative Cognition and Human Uniqueness." *Philosophical Transactions of the Royal Society B* 367, no. 1603 (2012): 2794–2802.

Sih, A., A. Bell, and J. C. Johnson. "Behavioral Syndromes: An Ecological and Evolutionary Overview." *Trends in Ecology and Evolution* 19, no. 7 (2004): 372–78.

Singer, P. "The Expanding Circle: Ethics and Sociobiology." New York: Farrar, Straus and Giroux, 1981.

Smith, A. *Wealth of Nations*. Wiley Online Library, 1999.

Smith, V. S. *Clean: A History of Personal Hygiene and Purity*. Oxford: Oxford University Press, 2007.

Sober, E., and D. S. Wilson. *Unto Others: The Evolution and Psychology of Unselfish Behavior*. Cambridge, MA: Harvard University Press, 1999.

Sommer, M. "Where the Education System and Women's Bodies Collide: The Social and Health Impact of Girls' Experiences of Menstruation and Schooling in Tanzania." *Journal of Adolescence* 33, no. 4 (2010): 521–29.

Soussignan, R., B. Schaal, L. Marlier, and T. Jiang. "Facial and Autonomic Responses to Biological and Artificial Olfactory Stimuli in Human Neonates: Re-examining Early Hedonic Discrimination of Odors." *Physiology and Behavior* 62, no. 4 (1997): 745–58.

Spencer, H. *The Data of Ethics*. Vol. 9. London: Williams and Norgate, 1887.

Sprengelmeyer, R., U. Schroeder, A. W. Young, and J. T. Epplen. "Disgust in Pre-Clinical Huntington's Disease: A Longitudinal Study." *Neuropsychologia* 44, no. 4 (2006): 518–33.

Spurier, M. F., M. S. Boyce, and B. F. J. Manly. "Effect of Parasites on Mate Choice of Captive Sage Grouse." In *Bird-Parasite Inter-*

actions: Ecology, Evolution, and Behaviour, edited by J. E. Loye and M. Zuk, 389–92. Oxford: Oxford University Press, 1991.

Stevenson, R. J., M. J. Oaten, T. I. Case, B. M. Repacholi, and P. Wagland. "Children's Response to Adult Disgust Elicitors: Development and Acquisition." *Developmental Psychology* 46, no. 1 (2010): 165.

Streidter, G. F. *Principles of Brain Evolution*. Sunderland, MA: Sinauer Associates, 2005.

Sugai, R., H. Shiga, S. Azami, T. Watanabe, H. Sadamoto, Y. Fujito, K. Lukowiak, and E. Ito. "Taste Discrimination in Conditioned Taste Aversion of the Pond Snail *Lymnaea stagnalis*." *Journal of Experimental Biology* 209, no. 5 (2006): 826–33.

Taylor, K. E. *Cruelty: Human Evil and the Human Brain*. Oxford: Oxford University Press, 2009.

Temple, S. "Do Predators Always Capture Substandard Individuals Disproportionately from Prey Populations?" *Ecology* 68, no. 3 (1987): 669–74.

Tolin, D. F., P. Worhunsky, and N. Maltby. "Sympathetic Magic in Contamination-Related OCD." *Journal of Behavior Therapy and Experimental Psychiatry* 35, no. 2 (2004): 193–205.

Tomasello, M., and M. Carpenter. "Shared Intentionality." *Developmental Science* 10, no. 1 (2007): 121–25.

Trivers, R. L. "The Evolution of Reciprocal Altruism." *Quarterly Review of Biology* 46 (1971): 35–57.

Troop, N. A., J. L. Treasure, and L. Serpell. "A Further Exploration of Disgust in Eating Disorders." *European Eating Disorders Review* 10, no. 3 (2002): 218–26.

Tybur, J. M., A. D. Bryan, D. Lieberman, A. E. Caldwell Hooper, and L. A. Merriman. "Sex Differences and Sex Similarities in Disgust Sensitivity." *Personality and Individual Differences* 51, no. 3 (2011): 343–48.

Tybur, J. M., D. Lieberman, and V. Griskevicius. "Microbes, Mating, and Morality: Individual Differences in Three Functional Domains of Disgust." *Journal of Personality and Social Psychology* 97, no. 1 (2009): 103.

Tybur, J. M., D. Lieberman, R. Kurzban, and P. DeScioli. "Disgust: Evolved Function and Structure." *Psychological Review* 120, no. 1 (2013): 65–84.

UNICEF and World Health Organization. "Diarrhoea: Why Children Are Still Dying and What Can Be Done." New York: UNICEF, 2009.

Weaver, J. E., and R. A. Sommers. "Life History and Habits of the Short-Tailed Cricket, *Anurogryllus muticus*, in Central Louisiana." *Annals of the Entomological Society of America* 62, no. 2 (1969): 337–42.

Weddle, C. B. "Effects of Ectoparasites on Nestling Body Mass in the House Sparrow." *Condor* 102, no. 3 (2000): 684–87.

Weiss, M. R. "Good Housekeeping: Why Do Shelter-Dwelling Caterpillars Fling Their Frass?" *Ecology Letters* 6, no. 4 (2003): 361–70.

West, M. J., and R. D. Alexander. "Sub-Social Behavior in a Burrowing Cricket *Anurogryllus muticus* (De Geer)." *Ohio Journal of Science* 63 (1963): 19–24.

Wheatley, T., and J. Haidt. "Hypnotic Disgust Makes Moral Judgments More Severe." *Psychological Science* 16, no. 10 (2005): 780–84.

Wilkinson, G. S. "Social Grooming in the Common Vampire Bat, *Desmodus rotundus*." *Animal Behaviour* 34, no. 6 (1986): 1880–89.

Wilkinson, S. I. *Votes and Violence: Electoral Competition and Ethnic Riots in India.* Cambridge: Cambridge University Press, 2006.

Wilson, E. O. *The Social Conquest of Earth.* New York: Liveright Publishing, 2012.

Wimberger, P. H. "The Use of Green Plant Material in Bird Nests to Avoid Ectoparasites." *Auk* 101, no. 3 (1984): 615–18.

Wolf, M., and F. J. Weissing. "Animal Personalities: Consequences for Ecology and Evolution." *Trends in Ecology and Evolution* 27, no. 8 (2012): 452–61.

Woody, S. R., C. McLean, and T. Klassen. "Disgust as a Motivator of Avoidance of Spiders." *Journal of Anxiety Disorders* 19, no. 4 (2005): 461–75.

World Health Organization and UNICEF. *Progress on Sanitation and Drinking-Water: 2010 Update*. Geneva: World Health Organization, 2010.

Wrangham, R. *Catching Fire: How Cooking Made Us Human*. London: Profile Books, 2010.

Wright, R. *The Moral Animal: Why We Are the Way We Are; The New Science of Evolutionary Psychology*. London: Vintage, 1995.

Wrubel, J., and S. Folkman. "What Informal Caregivers Actually Do: The Caregiving Skills of Partners of Men with AIDS." *AIDS Care* 9 (1997): 691–706.

Yan, Z., D. Ding, and L. Yan. "To Wash Your Body, or Purify Your Soul: Physical Cleansing Would Strengthen the Sense of High Moral Character." *Psychology* 2, no. 9 (2011): 992–97.

Yeshurun, Y., and N. Sobel. "An Odor Is Not Worth a Thousand Words: From Multidimensional Odors to Unidimensional Odor Objects." *Annual Review of Psychology* 61 (2010): 219–41.

Zhong, C. B., and K. Liljenquist. "Washing Away Your Sins: Threatened Morality and Physical Cleansing." *Science* 313, no. 5792 (2006): 1451–52.

Zimmer, C. *Parasite Rex*. New York: Touchstone, 2000.

Zito, M., S. Evans, and P. J. Weldon. "Owl Monkeys (*Aotus* spp.) Self-Anoint with Plants and Millipedes." *Folia Primatologica* 74, no. 3 (2003): 159–61.

INDEX